Stalin's Captive

Nikolaus Riehl and the Soviet Race for the Bomb

Nikolaus Riehl (1901–1990)

"A very nice man, expert in radiochemistry, former Director of Research of Auer-Gesellschaft." (Paul Rosbaud, the master spy, to Samuel A. Goudsmit in 1945.)

Stalin's Captive
Nikolaus Riehl and the Soviet Race for the Bomb

Nikolaus Riehl

Frederick Seitz
The Rockefeller University

History of Modern Chemical Sciences
Jeffrey L. Sturchio, SERIES EDITOR
Merck & Co., Inc.

1996
American Chemical Society
and the Chemical Heritage Foundation

Library of Congress Cataloging-in-Publication Data

Riehl, Nikolaus.
    Stalin's captive: Nikolaus Riehl and the Soviet race for the
bomb / Nikolaus Riehl: [introduction, commentary and translation by] Frederick Seitz.

    p. cm. —(History of modern chemical sciences)

Includes English translation of: Zehn Jahre im goldenen Käfig,
by Nikolaus Riehl.

ISBN 0–8412–3310–1

1. Nuclear bomb—Soviet union—History. 2. Riehl, Nikolaus—
Captivity, 1945–1955. 3. Physicists—Germany—Biography.

I. Seitz, Frederick, 1911–   . II. Riehl, Nikolaus, Zehn Jahre im
goldenen Käfig. English. III. Title. IV. Series.

QC773.3.S65R54   1995
355.8′25119′0947—dc20   95–43108
                                                          CIP

The paper used in this publication meets the minimum requirements of American National Standard
for Information Sciences—Permanence of Paper for Printed Library Materials, ANSI Z39.48–1984.

Copyright © 1996

American Chemical Society and Chemical Heritage Foundation

Series Foreword

Why study the history of chemistry? Chemical perspectives inform a diverse array of contemporary disciplines, from molecular biology and clinical pharmacology, to materials science and chemical engineering: Isn't keeping up with expanding research frontiers challenge enough? Indeed, in a world of email, fax, and on-line journals, it often seems we are drowning in a sea of information about chemistry in the present. Why, then, be concerned with chemistry in the past? The answer to that question offers intriguing perspectives on our understanding of the practices of the chemical sciences and technologies.

Chemists have always had an interest in the history of their science, and traditional accounts share certain common traits. As historian John Hedley Brooke has recently characterized the genre, these traits may be grouped under the headings of celebration, anticipation, autonomy, and continuity.[1] Historians of chemistry in earlier generations celebrated the achievements of their intellectual forebears, largely in relation to how well their theories and experiments anticipated later views. Traditional approaches also assumed a continuous progression of ideas from past to present, with developments following an internal logic largely divorced from broader intellectual, social, economic, and political currents. A splendid example of this mind-set comes from the work of the Swedish mineral chemist Torbern Bergman (1735–1784). In a 1779 essay entitled "Of the Origin of Chemistry," he divided its history into three distinct periods: "the mythologic, the obscure, and the certain."[2] This faith in the progress of chemical knowledge, culminating in the sophisticated present, is a leitmotif of much of the literature in the history of chemistry.

We find echoes of Bergman's periodization in nineteenth-century treatises like Thomas Thomson's *History of Chemistry* (London, 1830–1831) and Hermann Kopp's *Geschichte der Chemie* (Braunschweig, 1844–1847), as well as the work of more recent authors like J. R. Partington, whose *A History of Chemistry* (London and New York, 1961–1970) is still a standard reference. Similar sensibilities have motivated the generations of textbook authors who include the obligatory historical summaries of great discoveries, precursors, and heroic individuals in introductory general chemistry textbooks.[3] As interesting as these vignettes may be, chronicle is not history: The cumulative effect on students and scientists has often been a deadening of interest in historical inquiry.

But the history of chemistry offers other perspectives on the science, perspectives that help to place the quotidian activities of chemical scientists and engineers firmly in historical context. In recent decades, historians of science and technology have begun to elaborate on the myriad social, political, and economic dimensions of chemistry. The investigation of experimental practice, the study of scientific controversies, and the analysis of the reception and diffusion of new schools of thought illustrate the complex interplay of social and intellectual factors in interpreting theory change in chemistry. Attention to the relationship between chemical texts and their audiences are also illuminating the rich interactions of professional and public cultures in the development of chemistry. And there is increasing attention to invention and innovation in the chemical industries, the connections between academic and industrial research, and the reciprocal effects of chemical science and economic, business, and political structures. The result of a contextual approach to the history of chemistry is a fuller appreciation of the more complex interactions of science, technology, and society in the development of the "central science."[4]

A key goal of this new series on the "History of Modern Chemical Sciences," published jointly by the American Chemical Society and the Chemical Heritage Foundation, is to bring these new perspectives to a wider audience. Through an eclectic mix of original monographs, biographies, autobiographical memoirs, edited collections of essays and documentary sources, translations, classic reprints, and pictorial volumes, the new ACS–CHF series will document the individuals, ideas, institutions, and innovations that have created the modern chemical sciences. In so doing, we hope to foster among our readers a balanced view of the contributions that the chemical sciences and technologies have made to creating modern culture.

Edgar Fahs Smith (1854–1928), three-time President of the American Chemical Society and one of the pioneers of history of chemistry in America, expressed these sentiments in his last book, *Old Chemistries* (New York, 1927), in words that provide a fitting epigraph for our new series[5]:

> The criticism that chemistry is absolutely commercialized is frequently heard and, further, that it is the commercial value of the science alone which claims the thought of chemists. Such views are widely prevalent. But other ideas exist, and chemistry teachers especially know that to them the discarded "old chemistries" bring many other messages, messages in history, in philosophy, in economics, in social relations, in art, in international relations, in literature, and in a wide and extensive culture.

It is those "many other messages" about chemistry's past and future prospects that the ACS–CHF History of Modern Chemical Sciences Series is designed to explore. We welcome your thoughts on topics and directions for the series and, more importantly, your participation as an engaged audience.

Jeffrey L. Sturchio
Merck & Co., Inc.
July 1993

References

1. John Hedley Brooke, "Chemists in Their Contexts: Some Recent Trends in Historiography"; In *Atti del III° Convegno Nazionale di Storia e Fondamenti della Chimica*; Abbri, F.; Crispini, F., Eds.; Edizioni Brenner: Cosenza, Italy, 1991; pp 9–27, at 9–11.
2. I owe this reference to William B. Jensen, "Gibber, Jabber, or Just Geber?"; *Bulletin for the History of Chemistry* **1992**, *No. 12 (Fall)*, 49–50.
3. For a lively review of the Victorian tradition in history of chemistry, with useful commentary on other properties and uses of history of chemistry in the period, see Colin A. Russell, " 'Rude and Disgraceful Beginnings': A View of History of Chemistry from the Nineteenth Century"; *British Journal for the History of Science* **1988**, *21*, 273–294. *See also* Rachel Laudan, "Histories of the Sciences and Their Uses: A Review to 1913"; *History of Science* **1993**, *31*, 1-34; esp. pp 8–9, 16–17.
4. For an overview of some relevant recent studies, see Brooke, "Chemists in Their Contexts", esp. pp 16–24; and *Recent Developments in the History of Chemistry*; Russell, Colin A., Ed.; Royal Society of Chemistry: London, 1985. For a stimulating and perceptive guide to these trends in the history of science more generally, see Steven Shapin, "History of Science and Its Sociological Reconstructions"; *History of Science* **1982**, *20*, 157–211; and "Discipline and Bounding: The History and Sociology of Science as Seen Through the Externalism–Internalism Debate"; *History of Science* **1992**, *30*, 333–369.
5. Edgar Fahs Smith, *Old Chemistries*; McGraw-Hill Book Company: New York, 1927; p 89.

About the Author

Frederick Seitz was born in San Francisco on July 4, 1911. He received his Bachelor's degree from Stanford in 1932 and his Ph.D. from Princeton in 1934. He has written some classic works in physics including *Modern Theory of Solids* (1940), was co-editor of the series *Solid State Physics* (started in 1954), examined the evolution of science in *The Science Matrix* (1992), and wrote his autobiography in *On the Frontier: My Life in Science* (1994).

Seitz's early career included positions at the University of Pennsylvania, the Carnegie Institute of Technology, and General Electric. During World War II, he worked for the National Defense Research Committee, The Manhattan District, and as a consultant to the Secretary of War. From 1946 to 1947 he was director of the training program on peaceful uses of atomic energy at Oak Ridge National Laboratory. Appointed professor of physics at the University of Illinois in 1949, Seitz became department chair in 1957 and dean and vice president for research in 1964. He joined The Rockefeller University as its president in 1968.

Dr. Seitz was elected to the National Academy of Sciences in 1951, serving as part-time president for three years before assuming full-time responsibility in 1965. He has served as advisor to NATO, the President's Science Advisory Committee, the Office of Naval Research, the National Cancer Advisory Board, the Smithsonian Institution, and other national and international agencies. He has been honored with the Franklin Medal (1965), the Compton Medal—the highest award of the American Institute of Physics (1970), the National Medal of Science (1973), two NASA Public Service Awards (1969 and 1979), the National Science Foundation's Vannevar Bush Award (1983), as well as honorary degrees from 32 universities worldwide. In 1993, the University of Illinois renamed its Materials Research Laboratory in Dr. Seitz's honor. Stanford University has honored him with the Hoover Medal and Princeton University with the Madison Medal.

Contents

Photographs

Preface

It came as a surprise to me in 1993, while reading literature about wartime and postwar Germany and the Soviet Union, to learn that distinguished scientist Nikolaus Riehl (1901–1990), whom I had known about in the 1930s through his excellent published research, had spent 10 years as a scientist–captive in the Soviet Union. Initially, he worked on the production of pure uranium for the Soviet nuclear bomb program. On returning to West Germany in 1955, he wrote (in German) *Ten Years in a Golden Cage*, which was ultimately published by Riederer-Verlag in Stuttgart in 1988.

Finding a copy of his book was not easy because only a few copies had been printed, but I finally managed to acquire a photocopy through the friendly and generous support of his son-in-law and eldest daughter, Erich and Ingeborg Hahne of the University of Stuttgart, whom I later had the pleasure of meeting. In achieving this goal, I was aided substantially through good fortune by Annamarie Pick, the widow of Heinz Pick, formerly director of the Solid State Physics Institute of the University of Stuttgart, and H. C. Wolf of the Third Physical Institute of that university.

Riehl's account of his and his family's experiences in the Soviet Union between 1945 and 1955 is uniquely interesting in part because of its historical content; in part because he was bilingual in German and Russian, having grown up in St. Petersburg as the son of a German father and Russian mother; and additionally as a result of his warm human interest in the Russian people. Because the book was permanently out of print, I decided to translate it into English. I was aided in essential ways by Heinz and Mai von Foerster, former colleagues at the University of Illinois; the Hahnes; and an old friend, Leonard Lee Bacon—as well as by access to a good dictionary.

Along the way, I was able to gather important auxiliary information. I found an oral taped interview with Riehl that Mark Walker, a science historian at Union College in Schenectady, New York, had made. Transcripts of the interviews were generously made available by Carolyn Moseley of the Archive Center of the Niels Bohr Library of the American Institute of Physics at College Park, Maryland. She also provided me with access to the portion of the Samuel A. Goudsmit Files in the Archives that deal with the Alsos mission of the Manhattan District, for which Goudsmit was chief scientist. Similarly, a good friend, Alexander G. Bearn of The Rockefeller University, introduced me to Loren

Graham of the Massachusetts Institute of Technology (Cambridge), who loaned me a video tape of a 1988 *Nova* television program that is partly devoted to an interview with Riehl.

I owe Mark Walker a very special debt for many contributions to the accompanying manuscript. Not least, I appreciate our discussions of differing interpretations of events and personalities that appear in the text.

Two other individuals who, like Walker, had relatively direct relationships with Riehl after he returned to Germany are Zhores I. Alferov, the current director of the Ioffe Physico-Technical Institute in St. Petersburg, and Arnold Kramish, the author of the book on Paul Rosbaud cited in the introduction. An extract from a letter giving a brief account of Alferov's conversations with Riehl appears on page 195 of this book. I am also indebted to Alferov for a photograph of V. S. Emelianov.

While I was doing this research, the Farm Hall Papers were made public, providing additional information about the activities of some of the individuals with whom Riehl had worked intimately during the war. Then too, when I mentioned to Hans Mark of the University of Texas the name of the brilliant Austrian spy, Paul Rosbaud, who appears in the form of a helpful German colleague in the wartime saga, Mark informed me that Rosbaud had been a student of his father, the esteemed Herman Mark. He directed me to Arnold Kramish's book, *The Griffin*, which deals with Rosbaud's life and career.

Otto Westphal, formerly of the University of Freiburg, took a personal interest in the translation from the start and provided valuable advice along the way. Not least, he opened communication links with Manfred von Ardenne in Dresden and with Gustav V. R. Born, the son of the physicist Max Born and currently director of the William Harvey Research Institute of St. Bartholomew's Hospital Medical College in London. Westphal's introduction also led to a very interesting meeting with Gustav Born in New York City. In addition, I am indebted to Manfred von Ardenne for his interest in the translations and for providing me with a copy of a paper prepared in 1941 by Fritz G. Houtermans.

Fortunately, the book division of the American Chemical Society (ACS) decided, as I had hoped, that the translation, along with an introduction and annotations, had sufficient historical and human interest to merit publication in the form it appears here. Jeffrey L. Sturchio, editor of the ACS–Chemical Heritage Foundation History of Modern Chemical Science Series, helped me greatly in organizing the fragmentary auxiliary material I had assembled. I am indebted to Bruce Merrifield and Glenn T. Seaborg for their advice and generous support in my approach to the American Chemical Society, as well as for the superb guidance provided by the publishing staff.

Many individuals other than those already mentioned above were of help along the way. First and foremost was Florence Arwade, who maintained steadfast control over all aspects of the work from start to finish. I am also very grateful to the university librarian Patricia Mackey for special service.

Collecting the photographs was an adventure as is usually the case. Some are reproduced from a mint copy of the original book by Riehl that was generously loaned by Mark Walker, whom I came to know as a close neighbor at Lake George, New York. Others originated in the family photo collection of Ingeborg Hahne. I am grateful to her and her husband for duplicating them and allowing me to use them. Tracey Keifer, of the Emilio Segrè Visual Archives of the Bohr Library of the American Institute of Physics, was of great help in gaining access to the collection of photographs there. Sondra Bierre and her colleagues at the Hoover Institution Archives at Stanford University, California, provided much courteous assistance in obtaining photographs from their extensive files.

David Holloway, author of *Stalin and the Bomb* (Yale University Press: New Haven, CT, 1994), generously provided me with much advice, including that which led to the acquisition of special photographs from several sources, particularly the archives of the I. V. Kurchatov Memorial Institute in Moscow. There, Raisa Kuznetsova was superbly helpful. The same was true of Vladimir V. Belyakov of the Embassy of the Russian Federation in Washington, D. C., and Photo Novosty. Zhores A. Medvedev, in turn, provided me with photographs of his Russian geneticist mentors, Nikolai I. Vavilov and Nikolai V. Timofeyev-Ressovsky.

Charles and Martha Upton were of special support in numerous ways, particularly in connection with the search for pictures of Eugene Wigner and of Friedrich Paulus in relation to the Battle of Stalingrad. Heinz Maier-Liebnitz, who helped Nikolaus Riehl become established at the University of Munich after he arrived in West Germany, was highly cooperative in responding to my queries and providing me with a personal photograph.

Finally, I am grateful both to Riehl's wife, Ilse, and two daughters, Ingeborg Hahne and Irene Fiedler, as well as to Joachim Spencker of the Carl Hanser Verlag in Munich, which acquired the Riederer-Verlag, for permission to publish the translation of Riehl's book. Along the way, my lawyer–cousin, Ernst Gaus of Bruchsal near Heidelberg,

Germany, was of inestimable help to Riehl's family and me in clarifying the legal status of the original version of Riehl's book and in obtaining the necessary freedom to publish an English version.

This book could not have been written without the generous support of the administration and Board of Trustees of the Rockefeller University.

Translator's Note

I am a member of that generation of U.S. scientists that was expected to have a reasonable reading knowledge of scientific French and German by the time a Ph.D. was attained. In fact, I would estimate that more than half of the scientific reading I did before 1934, when I graduated from Princeton University Graduate School, was in those two languages. Moreover, much time spent in Europe since World War II and in following the European language periodicals has made it possible to retain continuing reading ability.

Nikolaus Riehl's *Ten Years in a Golden Cage* involves a more sophisticated and diverse German than one would normally find in most scientific works, at least in the fields in which I was engaged. Here a two-kilogram German–English dictionary (*Grosswörterbuch*, 2nd ed.; Harper-Collins-Klett: Glasgow, Scotland, 1991) was very useful because it contained many synonyms and figures of speech. One of my bilingual friends stated that Riehl employed an "elegant, classical German vocabulary"—one that has been much diluted since 1945 as a result of the presence of varying occupying forces and other influences.

The two closely linked taped interviews obtained by Mark Walker were much easier to translate because most individuals are more spontaneous in selecting words during speech than in writing with subsequent editing.

The Backdrop

Frederick Seitz

The preparation of this introduction stirred memories of many discussions held over some 60 years with the late Eugene Paul Wigner—chemist, physicist, mathematician, engineer, and beloved mentor for aspiring young scientists.

This small book tells part of the story, between 1939 and 1955, of the life of a remarkable, humane, sensitive scientist. Nikolaus Riehl found himself directing the manufacture of nuclear reactor grade uranium, first during World War II for two groups of German scientists, including that of Werner Heisenberg, who desired to produce a fission chain reaction and thereby (unknowingly) emulate Enrico Fermi's feat of December 1942, and then as a civilian scientist–prisoner in support of the Soviet scientists who were starting on the road to build a vast nuclear arsenal.[1] The Soviet espionage system must have been almost as effective in Germany as in the United States, although it operated differently. Riehl was picked up by Lavrenty Beria's Secret Police in May 1945 immediately after the Soviet army moved into Berlin.

The volume consists of two main parts. The first, this introduction, describes some of Riehl's life and work in Germany before becoming a scientist–captive in the Soviet Union and provides some details about the individuals with whom he was associated. It also gives a brief account of the experiences of members of the German Uranium Club during their incarceration at Farm Hall in England between May 1945 and January 1946 and describes segments of the careers of master Austrian spy Paul Rosbaud and Samuel A. Goudsmit, scientific head of the

---

[1]Detailed accounts of the overall Soviet Program on Nuclear Weapons appear in Stephen J. Zaloga's *Target America* (Presidio Press: Novato, CA, 1993) and David Holloway's *Stalin and the Bomb* (Yale University Press: New Haven, CT, 1994).

1

Alsos mission. The second part of the book is a direct translation from German into English, with footnote annotations, of Riehl's account of his period of captivity, *Zehn Jahre im Goldenen Käfig* (*Ten Years in a Golden Cage*). This book was published in Stuttgart in 1988 by Rieder-er-Verlag two years before Riehl died. The original manuscript, with subsequent additions, was written in the late 1960s and had been passed around privately within Riehl's circle of friends. Riederer-Verlag was acquired by the Carl Hanser Verlag of Munich soon after the first relatively limited printing, with the result that the book was difficult to obtain,

A 1988 *Nova* television program[2] calls Riehl a "Nazi scientist." This designation is entirely inappropriate. Riehl had nothing but disdain for the leaders of the National Socialist Party. Being part Jewish, he could not have condoned the racial laws that might have terminated his career. Indeed, his humane, balanced sense of values comes through clearly in the following material.

Why did individuals such as Riehl, who were not sympathetic to their government, not emigrate from Germany before the outbreak of World War II? A great worldwide depression was on, there were very few jobs abroad, and the jobs that were available went to a very select group of scientists who, almost without exception, had been forced out of Germany soon after Hitler took power in 1933.

Even the very capable Italian scientist, Edoardo Amaldi, who fervently wished to follow Fermi to the United States, failed to obtain a post because he was not a "true" refugee. He remained in Italy throughout the war. In an address to a group of scientists, which I attended in the early 1960s, Amaldi emphasized in the spirit of a warning that scientists, as citizens, are far from immune to political processes in their countries and must face that fact.

Someone such as Riehl, who was exceedingly talented in his field, might have been useful at the General Electric Company, RCA, or Westinghouse in the 1930s. However, those companies were not in a

---

[2]*Nazis and the Russian Bomb* deals with the role of some German scientists and engineers in the development of the first Soviet fission bombs and bomb-carrying rockets. Among those mentioned in Riehl's book, both he and the chemical engineer Gunther Wirths, his close colleague in Elektrostal, are featured. The brilliant inventor–scientist, Manfred von Ardenne, also appears. About half the documentary deals with the group from Peenemunde who helped in the initial stages of the Soviet rocket program. In the interview, Riehl emphasizes once again that the success of the Soviet programs was not critically dependent on the help of the German workers. He believes that this and other outside help speeded up the program by at most one or possibly two years.

position to hire a high-placed individual such as he, as I knew from personal experience as an employee of the General Electric Research Laboratory during the years of the Great Depression. Had he crossed the Atlantic before 1939, he could have expected no more than relatively menial work. Like many others, he stayed in Germany hoping for better days, which only came in the mid-1950s.

Werner Heisenberg's case is very different. He would have been welcomed in the United States or the United Kingdom. In fact, Columbia University (New York City) had an open offer to Heisenberg from 1937 onward, but family and national feeling, devoid of any admiration for Hitler, and a sense of loyalty to scientific colleagues who could not expect to find positions elsewhere, caused him to refuse this and other offers.[3] Such national feeling was common in the Europe of his time and still is to a degree in ours.[4]

Views about Heisenberg that stand somewhat in contrast to those expressed by Thomas Powers can be found in books by David Cassidy[5] and the science historian Mark Walker.[6] Cassidy attempts to grapple with the issue of the extent to which the German scientists in the Uranium Club might be said to have exhibited a higher level of morality

---

[3]See *Heisenberg's War* by Thomas Powers (Alfred A. Knopf: New York 1993, Chapter 1).

[4]Powers suggests that by the summer of 1939, during his last brief visit to the United States before the outbreak of war, Heisenberg was fully conscious that it might be remotely possible to produce a nuclear bomb under the right circumstances. He did not want to risk the chance of becoming involved in a program that would produce a bomb that might be used against his own country. Fermi, in turn, probably realized that there was much less chance that such a bomb would be used against Italy. Moreover, Fermi, being far more of a realist and having greater political insight, appreciated more fully the magnitude of the evils of which dictators are capable and the futility of scientists believing that they could serve as effective counterbalances. By 1938 when he left Italy, Fermi had been living under an ever-tightening dictatorship for 16 years. His plans for leaving had been solidified well before nuclear fission had been discovered.

[5]David Cassidy, *Uncertainty: The Life and Science of Werner Heisenberg* (W. H. Freeman: New York, 1991).

[6]Mark Walker, *German National Socialism and the Quest for Nuclear Power 1939–1949* (Cambridge University Press: New York, 1989); *Nazi Science, Myth, Truth and the German Atom Bomb* (Plenum Press: New York, 1995). A somewhat dramatized but well-studied version of the German nuclear reactor program is also given in *The Virus House* by David Irving (William Kimber: London, 1967). Irving consulted personally with many of the participants during preparation of the book.

than their American counterparts, as expressed by some Germans. He quite properly rejects the proposal.

Ten Years in a Golden Cage relates in warm, human detail the various adventures of Riehl and his family and colleagues during the 10 years in which he was a Soviet prisoner. He and his German group were released to go to Communist East Germany in 1955. Riehl, however, continued on to West Germany where he selected an academic career.

I became aware of the existence of his book while reading about the Soviet nuclear weapons program. I had known Riehl through his published research on crystal luminescence in the 1930s when I was working with the team at the General Electric Laboratory that developed the U.S. version of the fluorescent lamp Riehl had invented at the Auer Company in Berlin.

Riehl's Earlier Life and Times

Riehl was born in St. Petersburg in 1901 of a Russian mother and a German father. The latter was in charge of St. Petersburg operations of what is now known as the Siemens Electric Corporation. His mother's father was a well-known Jewish physician. The family lived in Russia through World War I and left for Berlin in 1918, after the new Communist government had made a separate peace with Germany.

The Riehl family was able to live on in St. Petersburg during World War I without being molested while Riehl's father carried on his professional work. This indicates that the Russian participation in World War I was focused less on what might be called a personal antipathy toward the Germans than on the desire of the Russian leaders to take over the Dardanelles for access to the Mediterranean; to restore Istanbul to Christian hands, as a holy mission; and to provide support to the Slavic communities within the Austrian Empire. Riehl's daughter, Ingeborg Hahne, states that during this period in Russia the family lived in a section of St. Petersburg where a number of other Germans, mainly professionals, had been welcomed before the war started and also suffered no ill consequences. Such communities of foreign specialists (often referred to as "German suburbs," although the population was usually mixed) had a long history in Russia. An account of the one in Moscow at the time of Peter the Great (1672–1725) appears in *Peter the Great* by Robert K. Massie (Ballantine Books: New York, 1980). See, for example, Chapter 9.

St. Petersburg, however, was the site of much revolutionary combat and other turmoil in 1918, with the result that the senior Riehl prudently decided that the family should move to Germany. Then, Riehl, who was in his late teens, was bilingual and had lived sufficiently close to the Russian people to have developed a deep personal affection for them—a relationship that appears clearly in his writings.

Once the family was settled in Berlin, Riehl decided to become a scientist, with an emphasis on nuclear physics and chemistry. In 1927 he obtained a doctorate in nuclear chemistry in Otto Hahn's Institute, with Lise Meitner[7] supervising his research.

While preparing to obtain the special diploma that would permit him to teach at a university, he accepted a position at the Auer Company in Berlin, which was devoted to the manufacture of special materials. For example, it extracted radium, mainly used in medical work and radiography, from uranium ore and refined the rare earths that were constituents of Welsbach mantles, commonly used to enhance the brightness of gas lamps. Although he acquired a teaching diploma, he did not use it until the 1950s because the opportunities at the Auer Company matched his talents closely. He was soon made director of advanced research, after which, in the 1930s, he invented the now ubiquitous fluorescent lamp while working in partnership with the Osram Lamp Company.

This great invention grew out of an interest he developed in the fundamental science associated with crystal fluorescence of the type observed in appropriately treated alkaline earth sulfides, such as zinc and cadmium sulfide. Once Riehl realized that the quantum efficiency of conversion from UV to visible light by such materials could be close to 100%, he decided that a highly efficient lamp could be produced by coating the inside of a tube containing a low-pressure mercury discharge with appropriate fluorescent crystalline powders. The overall energy efficiency of such a lamp for the production of visible light can attain the level of 25%—some 15 times greater than that of a typical hot-filament lamp of the time. The great value of this means of illumination was appreciated by the General Electric Company. General Electric's laboratory staff was soon working on an American counterpart of the lamp in close cooperation with the RCA Laboratories in Camden, New Jersey, where many new fluorescent compounds were being produced by H. W. Leverenz in support of Vladimir Zworykin's development of commercial television. During the early part of World War II, Riehl published a book that describes most of the essential work dealing with the fluorescent lamp on an international scale.

---

[7]An account of the career and travails of Lise Meitner is given by Ruth L. Sime, *American Journal of Physics* **1994**, 62, 695.

Lise Meitner, Riehl's thesis adviser at the Hahn-Meitner Institute. She and her nephew, Otto Frisch, played a catalytic role in the interpretation of Otto Hahn's discovery of nuclear fission in 1938. (Courtesy of the Herzfeld Collection of the Emilio Segrè Visual Archives of the American Institute of Physics.)

Otto Hahn. He and Riehl eventually became close friends. (Courtesy of
the Max Planck Institute and the Emilio Segrè Visual Archives of the
American Institute of Physics.)

It is perhaps worth adding that the fluorescent lamp was widely
adopted for use by industrial organizations in the United States during
World War II because of its high efficiency. This was especially the
case in new factories.

In 1984 and 1985, Mark Walker interviewed Riehl, then in his mid-
eighties, near Munich. Walker was working on his doctoral thesis for
the Department of History at Princeton University, was interested in the
history of German nuclear energy development between 1939 and
1949, and wished to gain knowledge of Riehl's experience in Germany.
These interviews took place about three years before the publication of
Riehl's memoir. Although some comments regarding Riehl's years in
the Soviet Union are made, they add little to what appears in his book.
The interviews took place in German and are available in the archives

Technische Physik

in Einzeldarstellungen

Herausgegeben von

W. Meißner, München und G. Holst, Eindhoven

3

Physik und technische Anwendungen der Lumineszenz

Von

Dr. phil. habil. Nikolaus Riehl

Direktor der wissenschaftlichen Hauptstelle der Auergesellschaft AG., Berlin

Mit 83 Abbildungen

Published and distributed in the Public Interest by Authority of the Alien Property Custodian under License No. A-266

Published by

J. W. EDWARDS

Lithoprinted by

EDWARDS BROTHERS, INC.

ANN ARBOR, MICHIGAN

1944

Berlin

Verlag von Julius Springer

1941

Title page of Riehl's 1941 book, summarizing the science of crystal luminescence and its applications. The book covers international developments, including his own innovative work, both basic and applied. The photograph is taken from a personal copy acquired in 1944 that had been reprinted through the Authority of the Alien Property Custodian.

of the American Institute of Physics.[8] Some of the results of Walker's research are published in the two books by him mentioned in footnote 6. His interviews with Riehl were accompanied by the hazard of asphyxiation: Riehl chain-smoked cigars during the entire exchange! Walker's oral history of Riehl forms the basis of the following section, which outlines relevant episodes in Riehl's career.

Incidentally, Riehl's account of the situation experienced by the scientists in wartime Germany shows how completely isolated they were—not only from the outside world but in a sense from one another. The available news was grossly distorted. One could trust only one's closest friends. The theoretical physicist J. Hans D. Jensen, who spent a major part of his career at the University of Heidelberg, once said: "We lived in a vacuum. We knew that all official news was riddled with falsehoods; we had no idea what was right."

Riehl's Reminiscences

As mentioned previously, Riehl's doctoral research was carried out under the immediate supervision of Lise Meitner. Although they never quarreled, the relationship between student and mentor was somewhat tense. Riehl found it difficult to work under the direction of a woman physicist, and Meitner was very strong minded and demanding; however, they did learn to accommodate each other.

Riehl became a close friend of Otto Hahn. They had similar open personalities and enjoyed conversing with one another on essentially a first-name basis—a relationship that not only persisted but grew steadily for many years because they were also neighbors in Berlin. Riehl was proud that Hahn, who was two decades older, came to refer to him as "my friend, Nikolaus."

Meitner, being Jewish on both sides of her family, was forced to flee Germany in 1938 after the unification with Austria, both Hitler's and her homeland. Her secret departure from Germany was arranged by Hahn and colleagues in Germany and Holland and is described in Hahn's book, *Otto Hahn: My Life* (English version, MacDonald: London, 1968). Hahn had first appealed to Carl Bosch, president of the Kaiser Wilhelm Society, for help in procuring an exit visa for Meitner, but Bosch was turned down by higher-ups.

During World War II, Riehl visited Meitner in Sweden, where she eventually settled. This was Riehl's only trip to a foreign country dur-

---

[8]The address of the Niels Bohr Library is One Physics Ellipse, College Park, MD 20740–3843; Phone (301) 209–3184; fax (301) 209–0882. I have carried out a free-flowing translation into English of the taped interviews, which is also available at the Institute.

ing the war. Meitner was very tense at the beginning of their meeting but relaxed somewhat as their discussions progressed and she realized that his intentions were those of a relatively grateful student who was paying homage to his mentor. Nevertheless, she expressed deep bitterness at what she regarded as betrayal by her former German friends.

Riehl retained his strong links with Hahn while at the Auer Company and closely followed the work of Hahn and Fritz Strassmann[9] as they went through the chemical and physical procedures that determined that fission actually occurred in uranium as a result of neutron capture. Riehl relates the special care and ingenuity that went into the work, because both of the leading investigators were highly experienced radiochemists.

Incidentally, the Auer Company, which was founded by the brilliant Austrian inventor Auer von Welsbach, was acquired by the Degussa firm soon after Hitler came to power. Previously, it had been under the control of a capable philanthropic Jewish entrepreneur whom Riehl calls Geheimrat Koppel, and for whom Riehl obviously had great respect and admiration. Leopold Koppel did much to develop the company and used a part of the profits during the difficult years following World War I to support basic scientific research in Germany in what are now the Max Planck Institutes. He was forced by Hitler's government to sell to the so-called "aryan" investors who controlled the Degussa firm in Frankfurt and then fled to Zurich for safety. Riehl was particularly bitter about this treatment of a very loyal and generous German citizen.

Riehl never openly disclosed his partial Jewish ancestry to the authorities, and the issue was never raised. He mentioned that Otto Hahn also had a Jewish antecedent but Hahn also never encountered any problems on this account. It is possible that Riehl was bypassed by those who were seeking out partly Jewish individuals, because he was working in industry. As a result of holding a high position in the Auer Company, he was able to shield within his organization partially Jewish colleagues by providing them with employment that did not carry managerial titles. He stated that toward the end of the war—a period in

---

[9]Accounts of the history of the discovery of fission may be found in W. R. Shea's (Editor) *Otto Hahn and the Rise of Nuclear Physics*, pp 91–133 (Dordrecht: Boston, MA, 1983). A suggestion by Aristide von Grosse, who was then at Columbia University, led Hahn to take up an investigation in depth. Enrico Fermi had obtained complex, inexplicable results with uranium when he carried out a systematic search of the products produced by neutron irradiation of many elements in the periodic system. (See the reproduction of a page from *Die Naturwissenschaften*, March 31, **1939**, *27*, 212 on page 12.) Von Grosse's role is acknowledged by Hahn in his autobiography (p 147).

Über den Nachweis und das Verhalten der bei der Bestrahlung des Urans mittels Neutronen entstehenden Erdalkalimetalle[1].

Von O. HAHN und F. STRASSMANN, Berlin-Dahlem.

In einer vor kurzem an dieser Stelle erschienenen vorläufigen Mitteilung[2] wurde angegeben, daß bei der Bestrahlung des Urans mittels Neutronen außer den von MEITNER, HAHN und STRASSMANN im einzelnen beschriebenen Trans-Uranen — den Elementen 93 bis 96 — noch eine ganze Anzahl anderer Umwandlungsprodukte entstehen, die ihre Bildung offensichtlich einem sukzessiven zweimaligen α-Strahlenzerfall des vorübergehend entstandenen Urans 239 verdanken. Durch einen solchen Zerfall muß aus dem Element mit der Kernladung 92 ein solches mit der Kernladung 88 entstehen, also ein Radium. In der genannten Mitteilung wurden in einem noch als vorläufig bezeichneten Zerfallsschema 3 derartiger isomerer Radiumisotope mit ungefähr geschätzten Halbwertszeiten und ihren Umwandlungsprodukten, nämlich drei isomeren Actiniumisotopen, angegeben, die ihrerseits offensichtlich in Thorisotope übergehen.

Zugleich wurde auf die zunächst unerwartete Beobachtung hingewiesen, daß diese unter α-Strahlenabspaltung über ein Thorium sich bildenden Radiumisotope nicht nur mit schnellen, sondern auch mit verlangsamten Neutronen entstehen.

Der Schluß, daß es sich bei den Anfangsgliedern dieser drei neuen isomeren Reihen um Radiumisotope handelt, wurde darauf begründet, daß diese Substanzen sich mit Bariumsalzen abscheiden lassen und alle Reaktionen zeigen, die dem Element Barium eigen sind. Alle anderen bekannten Elemente, angefangen von den Trans-Uranen über das Uran, Protactinium, Thorium bis zum Actinium haben andere chemische Eigenschaften als das Barium und lassen sich leicht von ihm trennen. Dasselbe trifft zu für die Elemente unterhalb Radium, also etwa Wismut, Blei, Polonium, Ekacäsium.

Es bleibt also, wenn man das Barium selbst außer Betracht läßt, nur das Radium übrig.

Im folgenden soll kurz die Abscheidung des Isotopengemisches und die Gewinnung der einzelnen Glieder beschrieben werden. Aus dem Aktivitätsverlauf der einzelnen Isotope ergibt sich ihre Halbwertszeit und lassen sich die daraus entstehenden Folgeprodukte ermitteln. Die letzteren werden in dieser Mitteilung aber im einzelnen noch nicht beschrieben, weil wegen der sehr komplexen Vorgänge — es handelt sich um mindestens 3, wahrscheinlich 4 Reihen mit je 3 Substanzen — die Halbwertszeiten aller Folgeprodukte bisher noch nicht erschöpfend festgestellt werden konnten.

Als Trägersubstanz für die „Radiumisotope" diente naturgemäß immer das Barium. Am nächstliegenden war die Fällung des Bariums als Bariumsulfat, das neben dem Chromat schwerstlösliche Bariumsalz. Nach früheren Erfahrungen und einigen Vorversuchen wurde aber von der Abscheidung der „Radiumisotope" mit Bariumsulfat abgesehen; denn diese Niederschläge reißen neben geringen Mengen Uran nicht unbeträchtliche Mengen von Actinium- und Thoriumisotopen mit, also auch die mutmaßlichen Umwandlungsprodukte der Radiumisotope, und erlauben daher keine Reindarstellung der Ausgangssalze. Statt der quantitativen, sehr oberflächenreichen Sulfatfällung wurde daher das in starker Salzsäure sehr schwer lösliche Ba-Chlorid als Fällungsmittel gewählt; eine Methode, die sich bestens bewährt hat.

Bei der energetisch nicht leicht zu verstehenden Bildung von Radiumisotopen aus Uran beim Beschießen mit langsamen Neutronen war eine besonders gründliche Bestimmung des chemischen Charakters der neu entstehenden künstlichen Radioelemente unerläßlich. Durch die Abtrennung einzelner analytischer Gruppen von Elementen aus der Lösung des bestrahlten Urans wurde außer der großen Gruppe der Transurane eine Aktivität stets bei den Erdalkalien (Trägersubstanz Ba), den seltenen Erden (Trägersubstanz La) und bei Elementen der vierten Gruppe des Periodischen Systems (Trägersubstanz Zr) gefunden. Eingehender untersucht wurden zunächst die Bariumfällungen, die offensichtlich die Anfangsglieder der beobachteten isomeren Reihen enthielten. Es soll gezeigt werden, daß Transurane, Uran, Protactinium, Thorium und Actinium

[1] Aus dem Kaiser Wilhelm-Institut für Chemie in Berlin-Dahlem. Eingegangen 22. Dezember 1938.
[2] O. HAHN u. F. STRASSMANN, Naturwiss. **26**, 756(1938).

*The first public announcement of the fission of uranium by Hahn and Strassmann as it appeared in a scientific journal. The title reads in effect, "On the Demonstration and Behavior of the Alkaline Earth Metals Generated by Irradiating Uranium with Neutrons." (Reproduced with permission from Die Naturwissenschaften **1939**, 27, 11. Copyright 1939 Springer-Verlag.)*

which there was great danger from reprisals at the hands of the Gestapo—he and his wife concealed in their home the Jewish wife of a senior employee. Individuals at any level of prominence faced great dangers on many scores during that period. This merciful action merely added one more dimension to the hazards of life at the time. Even Heisenberg was seriously threatened because of comments made while lecturing in Switzerland that were interpreted as defeatism. In this case, he was shielded through protective steps taken by Walter Gerlach, then Reich Plenipotentiary for Nuclear Physics in the Reich Research Council, which was by this time under the jurisdiction of Herman Göring.

Bemerkung zu den Untersuchungen von O. Hahn, L. Meitner und F. Straßmann über die Produkte, die bei der Bestrahlung von Uran mit Neutronen entstehen.

1934 beobachtete E. Fermi[1] bei der Bestrahlung von Uran mit Neutronen die Entstehung von mindestens 5 verschiedenen radioaktiven Atomarten. Anfangs untersuchte er nur eine davon mit der Halbwertszeit 13 Minuten näher. Er verglich einige chemische Eigenschaften dieser Atomart mit denen der Elemente, die im periodischen System dicht vor dem Uran stehen, da er zunächst annahm, daß eine Zertrümmerung des Urans unter ähnlichen wie den bisher bekannten Umständen, d. h. unter Aussendung von α- oder β-Teilchen, eingetreten sei. Die neue Atomart ließ sich jedoch mit keinem der bekannten Nachbarn des Urans chemisch identifizieren. Daraus und aus einer gewissen Ähnlichkeit einiger chemischer Reaktionen mit denen des Rheniums zog Fermi den Schluß, daß möglicherweise ein Element jenseits des Urans, und zwar wahrscheinlich ein Eka-Rhenium (Z = 93), entstanden sei.

Ohne die *Möglichkeit* der Entstehung von „Transuranen" aus dem Uran durch Neutronenbestrahlung zu bezweifeln, habe ich 1934 in einer kritischen Besprechung[2] der Fermischen Untersuchung betont, daß sein Ausschlußverfahren mir unvollkommen durchgeführt erscheine und daß Fermi sein neues Radioelement mit *allen* bekannten Elementen hätte vergleichen sollen. Ich sagte damals wörtlich:

„Man kann ebensogut annehmen, daß bei dieser neuartigen Kernzertrümmerung durch Neutronen erheblich andere „Kernreaktionen" stattfinden, als man bisher bei der Einwirkung von Protonen- und α-Strahlen auf Atomkerne beobachtet hat. Bei den letztgenannten Bestrahlungen findet man nur Kernumwandlungen unter Abgabe von Elektronen, Protonen und Heliumkernen, wodurch sich bei schweren Elementen die Masse der bestrahlten Atomkerne nur wenig ändert, da nahe benachbarte Elemente entstehen. Es wäre denkbar, daß bei der Beschießung schwerer Kerne mit Neutronen diese Kerne in mehrere *größere* Bruchstücke zerfallen, die zwar Isotope bekannter Elemente, aber nicht Nachbarn der bestrahlten Elemente sind."

O. Hahn und L. Meitner setzten Fermis Untersuchungen fort. Sie beschränkten sich zunächst auf den 13-Min.-Körper und einen zweiten von 90 Minuten Halbwertszeit; von beiden sagten sie in ihrer ersten Veröffentlichung[3] in Übereinstimmung mit Fermi[1], daß es sehr wahrscheinlich Elemente jenseits Uran seien. Sie nahmen wie Fermi von vornherein an, daß nur Elemente in der Nähe des Urans entstehen könnten.

Meine Kritik der Fermischen Versuche haben sie weder in ihrer ersten noch in einer ihrer vielen späteren Publikationen über die künstliche Umwandlung des Urans durch Neutronen zitiert. Mündlich auf diese Unterlassung aufmerksam gemacht, lehnte O. Hahn ein Zitieren meiner Arbeit ab, offenbar weil er meine Vermutung, daß das Uran vielleicht in größere Bruchstücke zerfallen könnte, für unsinnig hielt, da den Theoretikern damals solche Kernreaktionen unmöglich erschienen.

Wenn man die zahlreichen, in den nächsten 4 Jahren erschienenen Veröffentlichungen von Hahn, Meitner und Strassmann über die künstliche Umwandlung des Urans durch Neutronen und die dabei entstehenden Produkte sorgfältig liest, so fällt einem auf, daß die Verfasser, die immer weitere Produkte gefunden haben, ihre Ansichten über die untersuchten Prozesse und die entstandenen Atomarten vielfach änderten, daß sie etwas vorher „Bewiesenes" mehrfach zurücknahmen und wieder anders deuteten.

Am 18. November 1938[5] sagen Hahn und Strassmann zusammenfassend: „Als Folge der Neutronenbestrahlung der einen Atomart Uran 238 sind also bisher im ganzen 16 verschiedene künstliche Atomarten mit den Ordnungszahlen 88—90 und 92—96 nachgewiesen und in ihren Eigenschaften festgestellt worden." Sie hatten bis dahin 7 Transuran-Isotope (Z = 93—96) und je 3 Uran-, Radium- und Actiniumisotope festgestellt und beschrieben.

[1] Nature (Lond.) 133, 898 (1934).
[2] Angew. Chem. 47, 654 (1934).
[3] Naturwiss. 23, 37 (1935).
[4] Proc. roy. Soc. Lond. 146, 483 (1934).
[5] Naturwiss. 26, 755 (1938).

*An interesting paper by the physical chemist Ida Noddack, noting that in 1934, immediately after Fermi announced the complex results he had obtained with uranium, she had presented a paper in the journal Angewandte Chemie (**1934**, 47, 654) pointing out that the results obtained by Fermi indicate that the products of disintegration of uranium irradiated with neutrons involve a much more complete breakdown of the compound nucleus than the mere production of electrons, gamma rays, and alpha particles. Noddack states that she had informed Hahn of this possibility earlier in the decade but that he was then unprepared to accept such a complex idea. Apparently Hahn took Aristide von Grosse's suggestion more seriously (see footnote 9 on page 10). (Reproduced with permission from Die Naturwissenschaften **1939**, 27, 212. Copyright 1939 Springer-Verlag.)*

The German Uranium Club

As a result of his deep involvement in the development of the fluorescent lamp and other work at Auer during the second half of the 1930s, Riehl did not seem to appreciate the full significance of the discovery of fission—which he had been following—until he read a paper by theoretical physicist Siegfried Flügge, "Can One Make Technical Use of the Energy in the Atomic Nucleus?" (*Die Naturwissenschaften* **1939**, 27, 402). It was published several months before the start of World War II. Riehl realized that the Auer Company and its parent firm, Degussa, could become directly involved in the release of such energy through the production of pure uranium in the form of an oxide or a metal.

The Auer Company had a substantial store of "waste" uranium from which the radium had been removed. Riehl discussed the issue with Hahn after he noted the general interest in the topic that was developing among German physicists and was encouraged by Hahn to play a major role in helping those involved in experimental reactors. Hahn emphasized that his own interest lay entirely in the chemistry of the fission products. Riehl stated, however, that his own personal interest stemmed from three sources: scientific curiosity with respect to the possibility of developing a chain reaction, the long-range prospect of producing useful power and, not least, the opportunity to sell quantities of purified uranium, hitherto of almost no practical use. He also says that he would have dropped the issue had Hahn opposed the development of an experimental reactor at that time. Although Riehl is silent about whether he was fully aware of the potential uses of uranium as an explosive, he presumably had some degree of intimation of the possibility through discussions with colleagues.

In July 1939, Riehl went to the office of the Army Weapons Ministry to discuss the production of pure uranium. The officials expressed interest and provided him with an order to produce the oxide and the metal. Eventually, a special plant was built near Berlin for such production. Riehl gambled on investing company money in the new plant well before the orders were in hand. The key individual advising the ministry was physicist Kurt Diebner. Although the program was classified "secret," this classification was not always taken very seriously by the scientists involved. They sometimes clipped off the "secret" stamp from papers and letters and passed them around to interested scientists within Germany. They also discussed their early work with scientists from other countries, including Holland, Denmark, and Sweden.

The principal goal was to develop swimming pool reactors by using pure metallic uranium or uranium oxide, with heavy water as a moderator, although substitutes such as paraffin were used in the early stages of the work. Initially, the heavy water was purchased from Norway, but once that country was occupied, German engineers took over

Kann der Energieinhalt der Atomkerne technisch nutzbar gemacht werden?

Von S. FLÜGGE, Berlin-Dahlem*.

Zu Beginn dieses Jahres entdeckten HAHN und STRASSMANN[1], daß beim Beschießen von Uran mit schnellen oder langsamen Neutronen Barium, Lanthan und andere Elemente mittleren Atomgewichts entstehen. Die Entdeckung wurde sofort von zahlreichen Forschern in vielen Ländern aufgegriffen, und eine intensive Arbeit auf diesem Gebiet hat den Sachverhalt weitgehend geklärt und in mehr als 50 Veröffentlichungen schon zahlreiches quantitatives Material ergeben.

Im folgenden soll nur über ein Teilgebiet des ganzen, durch die HAHN-STRASSMANNsche Entdeckung angeschnittenen Fragenkomplexes berichtet werden. Gleich nachdem die Entdeckung der Zerspaltung von Urankernen sichergestellt war, wurde im HAHNschen Institut und wohl auch anderwärts die Frage aufgeworfen, ob bei einem so gewaltsamen Eingriff nicht auch einige Neutronen aus dem zerbrechenden Kern „abgedampft" oder „abgesplittert" werden könnten? Die Frage wurde auch alsbald in Angriff genommen, da sie zu einer sehr interessanten Konsequenz führte: Wenn jedes Neutron, das eine Aufspaltung hervorruft, im Gefolge der Aufspaltung 2 oder 3 Neutronen frei macht, so muß es möglich sein, daß diese Neutronen ihrerseits wiederum neue Aufspaltungen anderer Urankerne herbeiführen und auf diese Weise ihre Zahl noch weiter vergrößert wird, so daß eine Kettenreaktion ohne Ende schließlich zu einer Umsetzung des ganzen in dem bestrahlten Präparat vorhandenen Urans führen kann.

Man konnte dazu sofort einige Überlegungen anstellen, noch ehe man Einzelheiten kannte: Die Hauptfrage ist natürlich, ob und wie viele Neutronen je Spaltungsprozeß in Freiheit gesetzt werden. Dann kommt alles auf das weitere Schicksal dieser Neutronen an. Sie werden elastische Stöße ausführen können, die im wesentlichen nur ihre Richtung ändern; sie können unelastisch gestreut werden, so daß sie außer der Richtungsänderung auch noch eine beträchtliche Energieeinbuße erleiden; sie können eingefangen werden in der bekannten Reaktion

$$ {}^{238}_{92}U + {}^{1}_{0}n \longrightarrow {}^{239}_{92}U^{*} \xrightarrow[23^{m}]{\beta} {}^{239}_{}\text{Eka-Re}; \quad (1) $$

sie können endlich noch Einfangungen oder Umwandlungen an anderen Substanzen erleiden, die außer dem Uran anwesend sind, sofern man nicht reines Uranmetall bestrahlt, also z. B. am Sauerstoff von U_3O_8. Es wird darauf ankommen, ob all diese Reaktionen, welche nur Neutronen wegfangen ohne neue zu erzeugen, einen so großen Gesamtwirkungsquerschnitt haben, daß die beim Spaltungsprozeß erreichte Neutronenproduktion dadurch kompensiert wird oder nicht. Um zu erkennen, ob eine Kettenreaktion ablaufen kann,

müssen wir also über eine genaue Kenntnis aller konkurrierenden Wirkungsquerschnitte verfügen.

Endlich spielt noch eine dritte Frage eine große Rolle: die räumliche Ausdehnung der bestrahlten Substanzmenge. Die erzeugten Neutronen werden, ehe sie wieder einen Kern aufspalten, einen Weg von der Größenordnung einiger Zentimeter in der Substanz durchlaufen. Läuft also die Reaktionskette an einer Stelle der Substanz an, so breitet sie sich mit zunehmender Neutronenzahl über ein immer größeres Gebiet aus. Nun haben die Neutronen bei jedem elastischen Stoß die gleiche Chance zurückgeworfen zu werden, wie weiter nach außen zu laufen. Daher wird die Konzentration der freigesetzten Neutronen auch an der Ausgangsstelle der Reaktionskette zeitlich rasch ansteigen, sofern das benutzte Substanzvolumen so groß ist, daß der größte Teil der Neutronen oft zurückgeworfen wird, ohne die Oberfläche zu erreichen, durch die er die Substanz endgültig verlassen würde. Mit anderen Worten: Der Durchmesser einer bestrahlten Kugel aus uranhaltiger Substanz muß groß sein gegen die freie Weglänge, wird also einige Meter betragen müssen.

Ehe wir zur Diskussion der bisher angeschnittenen Einzelfragen übergehen, soll noch ein Wort gesagt werden über die Größenordnung der freiwerdenden Energie. Man kann sie leicht ungefähr abschätzen[2], ja sogar ziemlich genau angeben, daß jeder Spaltungsprozeß eine Energie von 180 MeV in Freiheit setzt[3]. Das läßt sich aus der Differenz der Massendefekte des Urankerns und der entstehenden Spaltungsprodukte herleiten[3]; die Zahl ist einigermaßen auch durch direkte Messung der kinetischen Energie der beiden entstehenden mittelschweren Kerne experimentell sichergestellt. Daß sich hierbei statt der erwarteten 180 MeV nur rund 160 MeV ergeben[4], kann schon als Hinweis darauf dienen, daß der Rest der Energie entweder noch in abgespaltene Neutronen gesteckt oder in Form von γ-Quanten abgestrahlt wird.

Der so erhaltene Energiebetrag ist sehr beträchtlich. Da die vorstehenden Überlegungen zeigen, daß es durchaus nicht ausgeschlossen ist, durch eine geeignete Versuchsanordnung eine Reaktionskette hervorzurufen, bei der das ganze Uran eines großen Blocks verbraucht wird, ist es zweckmäßig, sich einmal auszurechnen, wie groß z. B. die Energiemenge ist, die freigesetzt wird, wenn in 1 m³ U_3O_8 alles vorhandene Uran restlos umgewandelt wird. 1 m³ aufgeschüttetes U_3O_8-Pulver wiegt 4,2 t und enthält $3 \cdot 10^{27}$ Moleküle, also $9 \cdot 10^{27}$ Uranatome. Da je Atom etwa 180 MeV, d. h. rund $3 \cdot 10^{-4}$ erg oder $3 \cdot 10^{-12}$ mkg frei werden, wird insgesamt ein Energiebetrag von $27 \cdot 10^{15}$ mkg frei gesetzt, d. h. 1 m³ U_3O_8 genügt zur Aufbringung der Energie, welche nötig ist, um 1 km³ Wasser (Gewicht 10^{12} kg) 27 km hoch-

---

* Aus dem Kaiser Wilhelm-Institut für Chemie.

*Siegfried Flügge's speculative paper on the possibility of generating an energy-producing chain reaction by means of uranium fission. The title is "Can the Energy Content of Atomic Nuclei be Turned to Use?" (Reproduced with permission from Die Naturwissenschaften **1939**, 27, 402. Copyright 1939 Springer-Verlag.)*

Kurt Diebner, a scientist on the staff of the German Army Weapons Ministry. He not only helped Riehl procure funds to produce reactor-grade natural uranium but developed his own experimental reactor laboratory. He was interned at Farm Hall in England until January 1946. (Courtesy of Mark Walker and Farm Hall Transcripts.)

the production, which was interrupted by Allied attempts to sabotage the equipment. In November 1943, the Norsk Hydrodam at Rjukan, Norway, where the heavy water was produced, was destroyed by Allied bombing.

The research group guided by Fermi and Leo Szilard at Columbia University in New York and working in close cooperation with Eugene Wigner at Princeton University in New Jersey realized early in World War II that pure graphite could be a satisfactory moderator. The German groups did not pursue this route. Some say that they dropped carbon in preference to heavy water because they failed to evaluate the amount of boron in the test samples used to measure the capture cross sections. Walker states, however, that the issue of boron content was eventually well understood (see the first of his books mentioned in footnote 6 on page 3). Heavy water was selected, first because less uranium would be needed to achieve a critical structure and second to avoid the costly process of manufacturing pure graphite, which was not commercially available.

Two experimental programs were begun in Berlin. One was under the supervision of Kurt Diebner, within the authority of the Weapons Ministry. The other was in the hands of an academic group under the leadership of Werner Heisenberg and Karl Wirtz, a young experimental physicist. The first team used a lattice of cubes of uranium metal and oxide. The second used plates or sheets of uranium metal initially but then also used cubes. Riehl followed the work of both teams. Along with other physicists and chemists interested in the progress of the work, they formed what they designated the Uranium Club[10] (*Uranverein*).

Riehl expressed the view that the primary interest of the two groups, at least in the initial stages, was to see whether a chain reaction was feasible. None of the individuals with whom he dealt spoke either seriously or enthusiastically of developing a bomb, although the ultimate potential was recognized. Both of the programs were moved out of Berlin late in the war when the bombing became very heavy. Diebner's group was moved to Stadtilm in Thuringia, which eventually became part of East Germany, and the Heisenberg–Wirtz team moved to Haigerloch in Württemberg.

The two programs appear to have been both complementary and cooperative, if a little competitive because of the personalities of the individual leaders. Heisenberg, the leading theoretical physicist left in

---

[10]Until about 1950, the core disciplines of physics and chemistry, particularly inorganic, physical, and nuclear chemistry, were very closely linked. Leaders in the two fields not only possessed a great deal of mutual understanding but frequently focused their research on problems of common interest. A great change occurred in the post-war period when many physicists became interested in the properties of so-called high-energy particles with their ultimate implication for cosmology. At about the same time, many chemists turned their attention to the intricacies of biochemistry, with its ties to cellular and molecular biology. Fortunately, some core areas of science, such as those devoted to condensed matter, statistical mechanics, geoscience, and the properties of the atomic nucleus, continue to provide a common ground of interest for physicists and chemists.

Werner Heisenberg as he appeared in 1945 at the time of his internment at Farm Hall. (Courtesy of Mark Walker and Farm Hall Transcripts.)

Germany, wanted to see whether a chain reaction was possible.[11] Diebner's interests were similar, although he undoubtedly envisaged the early attainment of inexpensive power and presumably weapons in the long run.

The first mention in Germany of the possibility of producing a nuclear bomb appeared as loose talk by a few physicists soon after the discovery of fission, at a time when these individuals would have had little comprehension of the effort needed. Walker states (see footnote 6) that Heisenberg appreciated the potentialities of producing a bomb very early (1939) and so informed the military as he sought funds for

---

[11]Heisenberg published a book on cosmic ray physics, as well as papers on the theory of fundamental particles, during the war, indicating his attention was significantly divided by several interests during this period.

*Karl Wirtz, the young experimental physicist who worked with Heisen-
berg on the development of the planned experimental nuclear reactor.
This photograph, taken in his study, dates from a year or two after his
release from Farm Hall. (Courtesy of Mrs. Ottoni Wirtz.)*

reactor research. Friedrich C. F. von Weizsäcker, another, younger,
physicist member of the club, did much the same in 1940. Later, Heisenberg discouraged officials from pursuing any direct work on a bomb
because the resources needed were well beyond the national capabil-
ity during the course of the war.

The military ministry severely downgraded its interest in the program in 1942 when the German Blitzkrieg stalled at Stalingrad with
heavy losses. Industry became burdened with the production of more
conventional equipment; a great shortage of special materials, such as
copper, developed. Hitler and his advisers seized upon the opportunities that rocket development seemed to offer and focused their attention
on the V1 (buzz bomb) and V2 (liquid-fueled rocket) developed by
Wernher von Braun. Experimental work on the nuclear reactors was
turned over to a civilian science agency. This work was first under the
leadership of Abraham Esau, a respected technical physicist and influential National Socialist, and then in 1944 it was directed by Walter
Gerlach and had a very low priority. The production of reactor-grade
uranium proceeded very slowly. Riehl recounted the difficulties he had
during this period in obtaining the copper wire needed for a high-
temperature furnace.

The building at Gottow near Berlin where Kurt Diebner's research team housed its two experimental nuclear reactors. One device used cubes of uranium oxide embedded in paraffin, and the other used a similar cube suspended in heavy water. The sign states: "Entrance Forbidden." (S. A. Goudsmit photo, courtesy of the Emilio Segrè Visual Archives of the American Institute of Physics.)

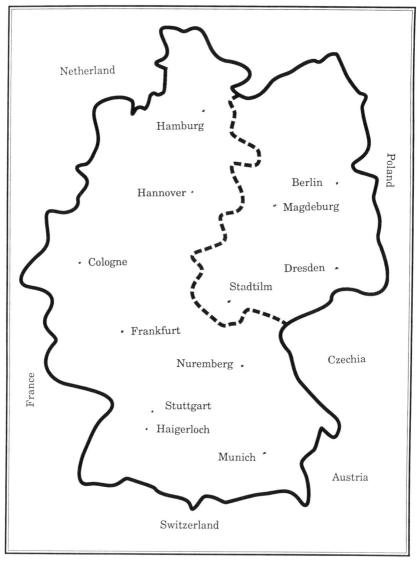

A schematic map of present-day Germany showing some principal cities as well as the locations of Haigerloch and Stadtilm, where, near the end of the war, the reactor components of the Heisenberg–Wirtz and Diebner teams were moved. The dashed line represents the border separating the former German Democratic Republic and the Federal Republic of Germany, commonly referred to as East and West Germany. The Western Allies had reached Berlin by April 1945 so there was ample time for the Alsos team to retrieve the portions of the reactor components that had not been moved to Bavaria by the Diebner team before the Soviets occupied all of the Eastern Zone during the summer.

Surprisingly, knowledge that the German nuclear program had been side-tracked in 1942 did not leak out to Switzerland, which presumably was teeming with spies. As we know, Heisenberg lectured in Zurich several times and encountered difficulties with the Gestapo because he once commented that Germany was losing the war. It appears that the actual status of the German uranium program was never discussed on these occasions, showing that some degree of secrecy was maintained outside Germany at that time. Paul Scherrer of the Swiss Federal Institute in Zurich, who knew Heisenberg well, was apparently in close touch with the U.S. Embassy during this period and would have reported anything of major consequence to it.

However, Niels Bohr had apparently received accurate information from Hans Jensen, the nuclear physicist at the University of Heidelberg, as well as others about the slow pace of the German nuclear reactor program but came to distrust it even though he admired Jensen and regarded him as a reliable friend. Jensen's family originated in Schleswig-Holstein, which had once been part of Denmark, and he felt close to the Danes. Presumably, the pace of the work underway at the Metallurgical Laboratory of the Manhattan District at the University of Chicago would have remained unaltered in any case, partly because of uncertainty about the accuracy of the information about the German program and partly because so much momentum had already been achieved.

On the more bizarre side, a U.S. intelligence team had posted an agent[12] (Morris "Moe" Berg) in Switzerland at the time of Heisenberg's visit to Zurich in 1944, with orders to assassinate Heisenberg if he mentioned anything about a nuclear bomb during a lecture. Heisenberg met the agent, who accompanied him to his hotel, and assumed that he was a Swiss student, perhaps the individual who later reported Heisenberg's comment about the war in a manner that allowed it to get back to the Gestapo (see the book by Thomas Powers in footnote 3 on page 3). After the war, Berg enjoyed socializing with the Alsos group, discussed later in this introduction.

Wartime at Auer and Human Problems of Political Origin

Only a small part of the overall work of the Auer Company during the war was concerned with the production of uranium. Most activity

---

[12]A colorful biography of Morris Berg, *The Catcher Was a Spy: The Mysterious Life of Moe Berg*, has been written by Nicholas Dawidoff (Pantheon Books: New York, 1994).

involved the manufacture of gas masks for the population at large—a relatively simple, profitable device that everyone was expected to have available in the event poison gas was used. Most of Riehl's activities centered around developments that could provide the company with new products "after the war." In retrospect, this was based on a highly simplistic view of the situation the company and its employees would face after the war, even though it may have appeared reasonable to the owners of the company at the time.

Riehl believed that most of the leaders of the government had a very low level of intellectual understanding of science and its importance for technology during Hitler's era. Moreover, he was inclined to believe that if Hitler had not come to power, Germany and the Soviet Union might have been able to work out a relationship of mutual advantage to both.[13] He felt that the Soviet leaders appreciated the opportunity of such a link and were "betrayed" by Hitler after signing a nonaggression pact. Because Riehl cannot be regarded as a deeply committed German nationalist, his support of this point of view probably rests on a combination of pragmatism and affectionate concern for the future of the Russian people, as expressed in his book. He regarded Stalin as far more intelligent and flexibly realistic than Hitler.

Before Hitler's rise to power, many Russian scientists published some of their best research either in the standard German journals or in a special German-language journal developed by the Springer publishing company that was specifically oriented toward papers originating from institutes in the Soviet Union. The journal was very popular and widely read and, because of this, many German as well as other foreign scientists were familiar with the scientists and the work that was being done in the Soviet Union.

Riehl mentioned that Gustav Hertz, who was Jewish on both sides of his family, had been shielded from the racial persecution of Hitler by the explicit actions of the Siemens family because he was an indispensable member of the company. Bitterness toward the persecution of the Jews by the government ran deep within Hertz, however, and in 1945 he more or less voluntarily went to the Soviet Union, where he put his best efforts into the development of a uranium isotope separation plant. He ultimately settled in East rather than West Germany, perhaps because it was under the strict control of the Soviet Union and yet

---

[13]This comment would seem to be in keeping with Riehl's concept of the importance of personal freedom only if one assumes that the link between the Weimar Republic and the Soviet Union would eventually have led to liberalization of the latter. That would scarcely have been conceivable before Stalin's death, because he had the basic ingredients of an absolute dictator.

A friendly professional meeting of Soviet and German scientists in Göttingen in pre-Hitler days (1926)—when wave mechanics was being formulated. From left to right are J. A. Krutov, V. J. Frenkel, S. I. Vavilov (later president of the Soviet Academy of Sciences, see photograph on page 130), M. Born, V. N. Kondratiev, P. Jordan, J. Franck, P. Kapitsa. (Courtesy of the V. Ia. Frenkel Collection of the Leningrad Physico Technical Institute and the Emilio Segrè Visual Archives of the American Institute of Physics.)

retained aspects of German living conditions with which he had grown up. Furthermore, he held almost unique celebratory status in East Germany of a type that was unlikely in the West. He also kept in East Germany, with Soviet support, many of the perks that he had been awarded in the Soviet Union. Riehl had disdained such matters almost completely, placing personal freedom at the top of his desires.

The case of Heinz Barwich, who was not Jewish but had grown up in a left-wing democratic political tradition, is more complicated than that of Gustav Hertz. He also went to the Soviet Union voluntarily, believing that its government represented the wave of the future. He was, however, bitterly disappointed by his experience. He ultimately

Gustav Hertz who, with James Franck, demonstrated the quantized nature of atomic excitation potentials. He was the nephew of Heinrich Hertz, who carried out the first experiments with electromagnetic radiation. I had the privilege of meeting him at the General Electric Laboratory in 1938 about the time this photograph was taken. (Courtesy of the Emilio Segrè Visual Archives of the American Institute of Physics.)

sought refuge in West Germany, shortly before his death, and with his wife wrote a book[14] describing his disappointment with the Soviet Union.

Another outstanding scientist who went to the Soviet Union voluntarily was the distinguished physical chemist Max Volmer, who had also been in Berlin at the end of the war. He was completely disgusted with experiences in Hitler's Germany and wanted to get away from it all. Riehl admired him very much and was planning to speak about him in a memorial session commemorating Volmer's one hundredth birthday that took place in East Berlin in the summer of 1985.

At the end of the war, it became clear that Nikolai V. Timofeyev-Ressovsky, who had been working in Berlin since the 1920s but remained a Soviet citizen, was in great danger. The Soviet authorities might accuse him of treason since he had remained in Germany during the Nazi period. As Riehl describes in his book, Timofeyev-Ressovsky was not molested by the Gestapo during the war but worked as a relatively free individual. Riehl tried to warn him of the hazards he faced, and in Riehl's interviews with Walker he provides somewhat more detail about his discussions with Timofeyev-Ressovsky than that in his book. He was completely unsuccessful, a matter for which Timofeyev-Ressovsky paid dearly by being sent to a gulag, where his health was undermined.

Riehl on Heisenberg and the Return to Freedom

As is to be expected, the somewhat enigmatic role played by Werner Heisenberg (see footnotes 5 and 6 on page 3) during the war emerged during Walker's interview of Riehl. Riehl regarded Heisenberg as a German patriot who was not at all enthusiastic about Hitler's government. He was, however, politically naive and, in trying to make the best of an increasingly bad, indeed hopeless, situation, made foolish comments to scientists in the occupied countries, implying it was better that Germany rather than the Soviet Union win the war. He apparently assumed that the Soviet Union would gain control of all of continental Europe if Germany were defeated, discounting the possibility of a successful invasion across the Channel from England.

Regarding fission research, Riehl emphasized his belief that Heisenberg, like many other German scientists, was primarily interested in the possibility of constructing an experimental, self-sustaining

[14]*Das rote Atom* by H. and E. Barwich (Scherz Verlag: Munich, Germany, 1967; Fischer Bücherei: Stuttgart, Germany, 1970).

The physical chemist Max Volmer, who voluntarily transferred his activities to the Soviet Union in 1945. This photograph was taken at the time of his seventieth birthday in 1955 just after he had returned to the East German part of Berlin. An honoring citation was presented by I.N. Stranski. (Courtesy of the Deutsche Akademie Leopoldina and the Deutsche Bunsen-Gesellschaft, Zeitschrift für Elektrochemie, Heft 5, Volume 59.)

nuclear reactor and not in developing a bomb. Actually, Heisenberg did little talking with German colleagues during the war other than in matters directly related to science. The dangers one faced from the Gestapo inhibited much open discussion.

Riehl stated that the knowledge that uranium of atomic mass 235 is a fissionable isotope was publicly known early on. In contrast, the knowledge that plutonium could be generated and was fissionable was held more closely but was not retained exclusively as a "private" matter within the Uranium Club. Once known, it was transmitted to the military authorities by von Weizsäcker in 1940. More explicitly, the potentialities that lay in plutonium were formulated in some detail in a paper by Friedrich G. Houtermans in August 1941. It demonstrates that the Uranium Club had an understanding of the existence and potential uses of plutonium. The full implications appear, however, to have been beyond the comprehension of the leaders in the German government. Houtermans was employed in Manfred von Ardenne's private laboratory, which was located in the section of Berlin that eventually became part of East Germany. In 1945 the Soviets moved it and all the equipment to the Soviet Union where it was re-assembled under von Ardenne's direction, as described in Riehl's book. Von Ardenne apparently went along fairly pragmatically with the shift in his program to the Soviet Union since he had no reasonable alternative and was given VIP treatment.

In his highly informative autobiography, *Die Erinnerungen* (Herbig: Munich, Germany, 1990), written when he was a citizen of East Germany, von Ardenne describes his great shock upon learning that the United States had used nuclear bombs in Japan; he had regarded the war in the Pacific as "essentially over" by August 1945. Perhaps in his mind, von Ardenne compared such use of the bombs with the 1945 fire-bombing of Dresden, a nonmilitary target, which actually took as many lives as either of the two bombs in Japan. He uses what he regards as the "irresponsible" employment of the nuclear bombs by the United States as part of his justification for his whole-hearted cooperation with the Soviet leaders in their own development of nuclear weapons (see Appendix A of this introduction).

At several points in his book Riehl confirms from his personal experiences many of the crueler, unjust aspects of Soviet life, which are described in Alexander Solzhenitsyn's books, particularly those in *The Gulag Archipelago*.[15]

When Riehl was asked whether he considered it remarkable that he possessed such good health as an octogenarian in view of the fact

---

[15]Alexander I. Solzhenitsyn, *The Gulag Archipelago* (Harper and Row: New York, 1974).

Zur Frage der Auslösung von Kern-Kettenreaktionen

von Fritz G. Houtermans

Mitteilung aus dem Laboratorium Manfred von Ardenne

Berlin-Lichterfelde-Ost

Inhalt

*The cover page of the unpublished 1941 paper by Friedrich G. Houter-
mans describing the existence and importance of the fissionable isotope
plutonium, which can be separated chemically from uranium. (Courtesy
of Professor Manfred von Ardenne.)*

Friedrich G. Houtermans, a brilliant, complex German physicist who conducted research in both Germany and the Soviet Union. Each country questioned his loyalty. (Courtesy of Manfred von Ardenne and Manfred von Ardenne, Erinnerungen: 65 Jahre für Forschung und Fortschritt by F. A. Herbig, Verlagsbuchhandlung GmbH: München, Germany.)

that he worked with large amounts of radioactive material during his early years at Auer and that he was a dedicated cigar smoker, he replied that it was in part the result of a fortunate, genetically determined, healthy constitution and partly a result of the great care he took to avoid undue exposure to radium and its emanations while at work. He had developed an almost inborn sense of caution during the period that he spent with Hahn and Meitner. He noted that Hahn was very

Manfred von Ardenne holding a version of an electron tube for television of his own invention. He was one of the leading innovators in the field. The book on the table is entitled Television Reception. *(Courtesy of Manfred von Ardenne.)*

fussy about Riehl's visits to the former's laboratory, not because of any fear for his personal health but because he was worried that Riehl would bring contaminating radioactivity into the laboratory after dealing with large quantities of such material at the factory. Riehl made it a practice to visit Hahn the first thing in the morning when fresh from his bath.

Start of a New Life

After returning to West Germany from the Soviet Union in 1955, Riehl wondered where to go, having decided that he would like to work with an academic institution. His old job at the Auer Company was gone because the company was now located in East Germany. A decision had been made under the Adenauer government to establish the main atomic energy center at Karlsruhe, near the Rhine River, but it was also agreed that the Technical University of Munich could develop a research reactor of the swimming pool type, which would be under the directorship of Professor Heinz Maier-Leibnitz. Heisenberg had been offered the directorship of the reactor, because he was then working at the Max Planck Institute for Physics in Munich with an associated appointment at the University of Munich, but he wanted nothing to do with it, presumably because the challenge of being the first to develop a nuclear chain reaction was gone.

Maier-Leibnitz invited Riehl to join him as a member of the reactor staff and he did so gladly. Riehl eventually gained a professorial position in which he limited his lecturing to special seminars. (This part of the Walker interview relates to matters freely available in the scientific and technical literature and hence need not be expanded upon here). His close German colleague, Gunther Wirths, who also returned to West Germany by choice, accepted employment at Degussa as an expert in the manufacture of reactor-grade uranium. Wirths, who is fluent in English, is featured in the Nova program cited in footnote 2.

Riehl was invited to spend a semester at New York University in New York City in 1965. He enjoyed the sabbatical in spite of the fact that he was not fluent in English and had also become hard of hearing. He describes some of the minor, amusing difficulties he experienced as a result, particularly in trying to understand the speech of a Texan.

Careers of Other German Scientists

Aspects of the careers of several German scientists, mainly chemists, who served in the Soviet Union, are described in a paper by Alfred

Heinz Maier-Leibnitz, the radiochemist who provided Riehl with the position on his experimental nuclear research team at the Technical University of Munich after the latter returned to West Germany in 1955. Maier-
Leibnitz spent a period in the United States as a research investigator
after World War II but returned to Munich. The photograph was taken in
the 1950s. (Courtesy of Heinz Maier-Leibnitz.)

Neubauer, derived in substantial part from the work of others,[16] that
was delivered at a conference in Veszprén, Hungary, in 1991. Riehl is
mentioned in the paper but only in passing.

---

[16]See Alfred Neubauer's "Im Hohlen Zahn des Löwen—Zum Wirken Deutscher
Chemiker in der Sowjetunion umittelbar nach dem II Weltkrieg." Report of the
First "Mineralkontor" International Conference on the History of Chemistry and
Chemical Industry, August, 12–16, 1991, pp 181–188 (separatum Technikatör-
ténëti Szemle XIX, 1992). Neubauer's report is based on studies made by U.
Albrecht, A. Heinemann-Grüder and A. Wellmann that have since been pub-
lished in the book Die Spezialisten (Dietz: Berlin, Germany, 1992).

Neubauer states that about 3000 German scientists and engineers, accompanied by about 3700 dependents, worked in the Soviet Union after the war. Most, about 36%, were involved in aircraft research and development, and 18% were rocket specialists. Some 3% were chemists. About 11%, including some chemists, were involved in the field of nuclear research. The tragedy for the Soviet people is that most of the effort supported the military rather than the civilian side of productivity.

Max Volmer chose initially to join Gustav Hertz at a research center on the Black Sea but had difficulty accepting orders from Hertz, the director of research and development, and returned to a laboratory of his own in Moscow in 1946. He was followed to Moscow by several of his former colleagues who had also been with him on the Black Sea. On arriving in the Soviet Union in 1945, Volmer had been assured by the Soviet leaders that he could return to Germany in eight years, that is 1953, at which time he would be 68 years old. When this promise was broken for reasons of military security, he stopped work and awaited a decision that would allow him to leave. This occurred early in 1955, at which time he returned to East Germany and served as president of the East German Academy of Sciences for several years before going into full retirement.

Von Ardenne remained in Sinop near Sukhumi, working on the magnetic separation of isotopes until he shifted his interest, at the sug-

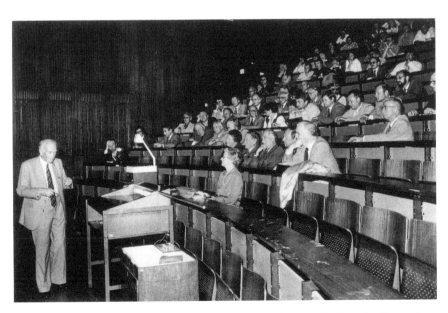

Nikolaus Riehl delivering a lecture in the conference hall at the Technical University of Munich in about 1960. (Courtesy of the Riehl family.)

gestion of the Soviet authorities, to the study of plasma research that might lead to controlled fusion. With the new assignment, he made extended trips to various research centers, including some in Moscow and Leningrad. He returned to East Germany in 1955 and established a new private institute in Dresden with the help of the Soviet and East German governments.

Peter Thiessen joined Hertz in Agudseri near Sukhumi in a program devoted to the separation of the uranium isotopes by using gaseous diffusion and uranium fluoride. Hertz, who had developed a similar process for other isotopes in the 1920s, had a good deal of experience in this procedure. One of Thiessen's principal responsibilities was to develop porous metal tubing made of nickel to serve as a diffusion barrier. The process developed proved to be so successful that in 1948 Thiessen moved part-time to Kyshtym with his colleague Heinz Barwich. This location, in the heart of the industrialized district east of the Ural mountains, was much more practical for establishing a major production plant for enriching uranium 235.

The group involved in this area of development had produced sufficient enriched material by 1951 that it proved possible to test a pure uranium bomb. The first Soviet bomb, tested in 1949, was based on the use of plutonium—the program in which Riehl had been involved. In 1952 Thiessen moved to Elektrostal near Moscow, where Riehl was first located, and headed a large group involved in technical problems associated with nuclear energy. He returned to East Germany in December 1956 to head an institute of the academy of sciences devoted to physical chemistry.

As is mentioned in Riehl's book, Hertz moved to Moscow in 1952 and remained in the Soviet Union until 1954 when he was made head of an institute of physics in Leipzig, East Germany.

Alsos, Heisenberg, and the German Nuclear Bomb Project

As the Alsos intelligence team created by General Leslie R. Groves, the head of the Manhattan District, that is, the entire wartime nuclear fission program, moved into northern Europe after the successful Normandy Invasion in 1944, it had two major missions: to learn how far the Germans had gone in the exploitation of nuclear energy, for bombs or otherwise, and to interrogate the principal individuals involved in the work. The immediate search was led by Boris Pash, formerly of the Federal Bureau of Investigation, and Samuel A. Goudsmit,[17] a distinguished physicist from the University of Michigan and a native of Hol-

---

[17]See S. A. Goudsmit's Alsos (Henry Schuman: New York, 1947).

*Peter A. Thiessen, a colloid chemist who had been head of the Kaiser Wil-
helm Institute for physical chemistry in Berlin until 1945 and who voluntarily followed von Ardenne to the Soviet Union and then joined Hertz's
group there. (Courtesy of Professor von Ardenne and Manfred von
Ardenne, Erinnerungen; 65 Jahre für Forschung und Fortschritt by F. A.
Herbig, Verlagsbuchhandlung GmbH: München, Germany.)*

land, who had been involved in work with the Radiation Laboratory of
the Massachusetts Institute of Technology both in Cambridge, Massachusetts, and in England. The first of these quests revealed to their
astonishment how little in the way of practical results the German scientists had achieved after six years of work relative to developments in
the United States, carried out in cooperation with the scientists of the
United Kingdom. In brief, the German work was parallel to that of the
United States only until mid-1942.

The second led to the interrogation of many German scientists, with
the ultimate retention and incarceration of 10 at a country estate—

Boris Pash, head of the Alsos mission, in action in 1945 near Haigerloch in Bavaria where the Heisenberg–Wirtz reactor had been moved. (Courtesy of the Emilio Segrè Visual Archives of the American Institute of Physics.)

Samuel A. Goudsmit just at the end of the happier, prewar days (July 23, 1939). He is with a group of scientist colleagues at a conference at the University of Michigan in Ann Arbor. From left to right are Samuel Goudsmit, Clarence Yoakum of the Graduate School, Werner Heisenberg, Enrico Fermi, Dean Edward Kraus. Goudsmit and Heisenberg were still warm friends at that time, although tensions were building because colleagues at the meeting were urging the latter to emigrate immediately to the United States. Fermi actually had left Italy permanently for the United States at the end of 1938. (Courtesy of the Crane Randall Collection of the Emilio Segrè Visual Archives of the American Institute of Physics. The photograph was taken by Dean Kraus' son John, later a professor of physics at the University of Ohio.)

designated Farm Hall—at Godmanchester near Cambridge, England. These scientists were held incommunicado for approximately eight months beginning in May 1945, soon after the end of the European phase of the war, until January 1946. Most of the names of those detained appear in Riehl's book or the Walker interviews. They are Erich Bagge, Kurt Diebner, Walther Gerlach, Otto Hahn, Paul Harteck, Werner Heisenberg, Horst Korsching, Max von Laue, Friedrich C. F. von Weizsäcker, and Karl Wirtz. Korsching, the least well-known member of the group, was an assistant at the Kaiser Wilhelm Institute for Physics, where Heisenberg became director in 1942. Although the captives were visited by several British scientists such as Patrick Blackett, Charles Frank, and Charles D. Darwin and had access to standard books and newspapers, they were not permitted to communicate with their families in occupied Germany until late autumn.

Walter Gerlach of the famed Stern–Gerlach Experiment of 1923. In 1944
he took over the directorship of the organization supporting basic scientific research in nuclear physics in Germany. The work on experimental
nuclear reactors had been transferred to the jurisdiction of a civilan
agency in 1942 with a low priority for essential equipment. This photo
was taken during his internment at Farm Hall. (Courtesy of Mark Walker
and the Farm Hall Transcripts.)

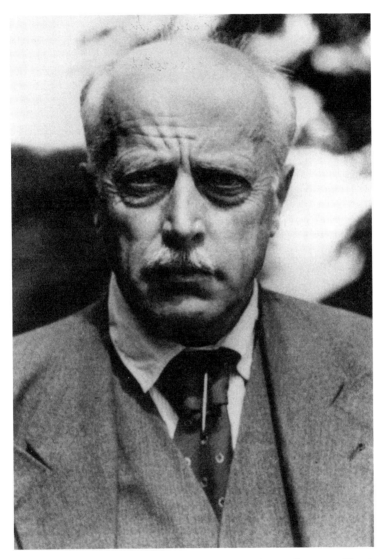

Max von Laue, who discovered the diffraction of X-rays by crystal lattices in 1913 and opened the door to both X-ray spectroscopy and the determination of crystal and molecular structures. He was, by choice, not part of the Uranium Club but was nevertheless interned at Farm Hall. This photograph was taken at that time. (Courtesy of Mark Walker and the Farm Hall Transcripts.)

F. Charles Frank, who played a major role in the interrogation of the German scientists detained at Farm Hall. The photograph dates from the mid-1980s. (Courtesy of the Physics Today *collection of the Emilio Segrè Visual Archives of the American Institute of Physics.)*

At Farm Hall the conversations of the group were monitored and recorded surreptitiously by a German-speaking staff and translated into English. The English transcriptions came into the public domain in 1992 and are readily available in book form, with descriptions of the course of their careers following detention.[18] The German originals apparently were not retained. It is not known to what extent the incar-

---

[18]The version of the transcripts that was stored in the U.S. National Archives has been published by the University of California Press (Berkeley, CA, 1993) under the title *Operation Epsilon: The Farm Hall Transcripts*, with an introduction by F. Charles Frank, who was involved in the creation of the project. The volume includes annotations as well as figures. A German translation is also available: *Die Farm-Hall Protokolle oder die Angst der Allierten vor der Deutschen Atombombe* by Dieter Hoffman (Rowohlt Berlin: Berlin, Germany, 1993).

cerated group suspected that their conversations were being recorded, although Karl Wirtz expressed suspicions to Charles Frank late in the cycle of events. The direct evidence, reinforced by a statement by Werner Heisenberg, the leading scientist in the group, shows that they did not realize their conversations were being recorded for most of their internment. Taken as a whole, the Farm Hall record has what might be called an iridescent quality in the sense that it can be viewed in many lights and given many interpretations depending on the knowledge and emotional attitude of the reader, and on the aspects of the transcripts that are deemed most significant.

Several points should be made at the start of any discussion of the transcripts. When Goudsmit, who had known Heisenberg for many years, first met with him early in May 1945, he apparently told him, on being asked, that there had been little follow-up work on uranium fission in the United States during the war—which was not true. Goudsmit's intent apparently was, in part, to try to convince the German group that their own failed research represented the leading edge of such work.

However, the first successful, and secret, test of the bomb at Alamogordo, New Mexico (July 16, 1945), actually took place two months after this conversation, whereas the Hiroshima and Nagasaki bombings, accompanied by much publicity, took place early in August nearly three months later. In fact, the development of working bombs was still in limbo to a considerable extent in May, so that in a sense Goudsmit was not entirely devious, granting that some 600 megawatts of nuclear energy were being generated in the nuclear reactors at Hanford, Washington, at the time.

As a result of the false assurances of Goudsmit, along with a well-nurtured strain of arrogance linked to over-confidence, the German group expressed disbelief that the bombs dropped on Japan were nuclear when first announced in the press in early August. This view was shattered in stages during the next few days as further details, both accurate and distorted, emerged through the media. The psychological reaction of members of the German group was varied, depending on the previous role and temperament of each. Otto Hahn, the discoverer of fission, became morbidly depressed for a period, experiencing a deep feeling of guilt for being the discoverer of nuclear fission. Walter Gerlach, who was nominally in charge of the German program in its later civilian stage, expressed personal anguish and self-recrimination for lack of success in achieving a chain reaction.

Heisenberg, after sharing the initial doubts, moved on with remarkable speed. He soon emerged from the initial traumatic stage of disbelief and began to reconstruct, as well as was possible under the limited circumstances and with the aid of his memory of the appropriate experimental parameters, the way in which a primitive "gun-type " bomb, composed of two hemispheres of enriched uranium of mass 235, could operate. His lecture, given to the group on August 14, eight days after the first announcements, was remarkably succinct and well-directed, in spite of the somewhat irrelevant interruptions by his partly confused companions. If one can assume that he actually had not given thought to the design of such a bomb before being incarcerated, his lecture provided a clear and not surprising reflection of his remarkable insight into physical problems.

In purely human terms, the most striking feature of the transcripts is the high level of anxiety felt by most members of the group. Those with families had been taken away somewhat vindictively by foreign military personnel at a time when German society was in a complete state of collapse. The captives were retained under conditions in which they had no direct way of learning the fate of wives, parents, and children. Moreover, they had no way of knowing what their own fate would be.

Furthermore, the group of 10, although for the most part linked through the Uranium Club, were quite dissimilar in many other respects. As a result, internal tensions ran high at times. Even under the best circumstances, the group would not have formed a congenial fraternity. However, the level of anxiety and tension within the group diminished rapidly when, toward the end of 1945, they received news that they would be released early the following year. They were even willing to express considerable gratitude to their captors for what they then recognized had been privileged treatment under very difficult circumstances.

Apart from scientific, technical, and psychological issues, there are two ethical matters raised by the Farm Hall transcripts. First, was the incarceration necessary and appropriate under the circumstances in which it occurred, leaving out details of the treatment of the individuals and what might be termed "legal" issues? Here I am inclined to take a definitely affirmative view. As a result of the failure of the intelligence organizations of the United States and the United Kingdom, or perhaps because of failure of communications between the two countries, the Western Allies did not know how far the Germans had gone in the exploitation of nuclear energy during the war. From personal experience, I can say that the atmosphere at the Metallurgical Laboratory of the Manhattan District at the University of Chicago was clouded with great uncertainty on this issue. In some individuals, such as Eugene Wigner, the level of concern regarding possible German progress bordered on panic.[19] A complete site investigation in Germany was

_____

[19]In contrast, Hans Bethe, appreciating the limited intellectual understanding of the German political leaders and the undoubtedly strained economy of the country, was much less pessimistic about the threat of a German nuclear bomb.

required as soon as circumstances permitted. It would not be possible to accomplish this without questioning the principal Germans involved in the research and thereby revealing the Western Allies' interest in nuclear energy before the program of the Manhattan District had reached fruition. Therefore, it was necessary to conduct the investigation in some form of isolation.

Perhaps isolating the scientists was less justified after the bombs had been used in Japan and the war in the Pacific had ended. However, the European situation was still very fluid at the end of the summer of 1945. In particular, it was not known to what extent the leaders of the Soviet Union were prepared to continue to work in partnership with the West toward common goals in the period of peace that lay ahead. Furthermore, there undoubtedly was a desire on the part of some individuals to witness the reaction of the German scientists to the success of the Western program in nuclear energy. Great anger over the enormous crimes committed by German political leaders and the desire to "rub it in" to the populace, including the scientists, were prevalent. Minor mistakes were made, such as the inclusion in the group of Max von Laue, who was not a member of the Uranium Club, but no significant harm was done.

In conclusion, the incarceration at Farm Hall and the events associated with it, embarrassing and traumatic though they may have been to the captives, were completely justified by the unusual circumstances surrounding the war and the failure of the intelligence network to provide accurate information to the Western Allies regarding the status of the German program. In a sense, the procedure followed can be supported ethically on the basis of what might be termed a force majeure—that is, the need to resolve a major issue in a way that can not be handled in a simple alternative manner. It is significant that the complaints of the German scientists abated once their ultimate fate was known to them.

The second ethical issue is far more complex and centers around the way one evaluates the wartime attempt of the German scientists to achieve a nuclear chain reaction. Here we move into a gray area in which the mindset of the analyst or judge is probably just as relevant as the actions of the individuals in the German Uranium Club. As Nikolaus Riehl states so clearly in his book and interviews, it would have been very difficult for a nuclear scientist or an engineer not to wonder whether a chain reaction could be achieved and not to seek funds from whatever sources were available to carry out experimental tests. Any major criticism of the key scientists and engineers can scarcely be based on their desire to satisfy that very natural curiosity.

Criticism must stem from more subjective factors such as one's opinion concerning the extent to which one or more members of the

Eugene P. Wigner, an early, major contributor to the formalization of wave mechanics. He was a Hungarian by birth and a refugee from Hitler. With a small research team, he took the lead in developing detailed plans for large-scale nuclear reactors in the United States during and after World War II. He obtained his university education, as well as early academic posts, in Germany. His initial training was in chemical engineering, but his deepest interests were in theoretical chemistry and physics. His familiarity with the great potentialities of German industry caused him to fear the early development of nuclear weapons in Germany. (Courtesy of Princeton University.)

group actually desired to aid the political leaders in the pursuit of the war—a difficult matter to ascertain under the circumstances that occurred. Such ethical judgments would have been far easier to make if the group had actually succeeded in achieving an operating nuclear reactor and had then become involved in specific applications, military or otherwise. Because the group failed in its primary goal, judgments must inevitably rest on one's interpretation of statements made by individuals or, less tangibly, "attitudes." Here, emotional factors come into play on both sides of the issue, granting that emotional factors do have a place in human affairs. As it stands, the subject is left open to endless discussion and interpretation.

Rosbaud, the Russians, and Riehl

"A very nice man, expert in radiochemistry, former director of research of Auer-Gesellschaft." (Paul Rosbaud's comment on Nikolaus Riehl.)[20]

One may speculate on the means by which the Russians became familiar with the work of the German Uranium Club well before the war ended—information which caused them to seek out Nikolaus Riehl almost as soon as they swarmed into Berlin in 1945. One not unlikely route is that provided by the brilliant espionage activity of Paul Rosbaud.

I first heard the name of Paul Rosbaud while having lunch with Samuel Goudsmit and Howard P. Robertson[21] in the Officer's Mess at Allied Military Headquarters in Frankfurt, Germany, during the summer of 1945. This occurred soon after the end of the war in Europe and soon after our technical intelligence office, under the direction of Robertson and part of General Eisenhower's staff, had moved from Versailles to Frankfurt.[22] Goudsmit, as scientific head of the Alsos team, planned to meet with Rosbaud. He stated that Rosbaud had been one of the most remarkable Allied espionage agents in Germany during

---

[20]Paul Rosbaud, the master spy, to Samuel A. Goudsmit in a letter of September 16, 1945. (Box 1, Folder 25, in the Alsos portion of Goudsmit Papers held by the Niels Bohr Library of the American Institute of Physics: College Park, MD. See Folder 41-47).

[21]H. P. Robertson's scientific career is described in the memoir prepared by Jesse L. Greenstein, *Biographical Memoirs of the National Academy of Sciences* (National Academy of Sciences Press: Washington, DC, 1980, Vol. 51, p 343).

[22]For additional details, *see* F. Seitz's *On The Frontier, My Life in Science* (American Institute of Physics Press: Woodbury, NY, 1994).

Paul Rosbaud, the Austrian scientist who served as a key member of the staff of the Springer Publishing Company before and during World War II. He used his post to establish a broad espionage network that relayed important information to the British intelligence service. This photograph was taken in New York City in 1961 when he became the first recipient of the Karl Taylor Compton Medal of the American Institute of Physics. (Courtesy of the Physics Today collection of the American Institute of Physics.)

the war, particularly with respect to scientific and technical information, although the range of his valuable work was much broader. Robertson, from the United States, had spent most of the war in England as part of a high-level bilateral intelligence group.

Rosbaud's key links, so far as is known, were primarily with British and not American intelligence agencies. Only the barest outline of his detailed activities and methods seems to be available in the open literature: partly because few records were kept and partly because he and the British wished it to remain so. Arnold Kramish, a popularizer of history, has attempted to piece the story together in a somewhat novel-like biography.[23] What seems certain is that through his numerous contacts Rosbaud collected and transmitted an enormous amount of information with much stealth and cleverness. He operated in Germany, occupied France, and Scandinavia and always at great risk to himself. His combination of perceptions and adroitness is, in its own way, somewhat reminiscent of that of the French nobleman Talleyrand, who maneuvered on the diplomatic front in France during the turbulent years that began with the start of the French Revolution in 1789. Rosbaud, like Talleyrand, was one of those who looked far ahead and was well prepared as events progressed.

According to Kramish's studies, Rosbaud was born as an illegitimate child of known parentage in Graz, Austria, in 1898. He served in the Austrian army on the Italian front in World War I and was taken prisoner. He became familiar with individuals in the British army during his period of captivity and developed a great admiration for the British.

He attended the technical university in Berlin after the war, obtaining a doctoral degree under the supervision of Herman Mark— the brilliant polymer chemist who spent his later career as a distinguished refugee in the United States—with whom he developed a close friendship. His first position was with a large metallurgical firm in Frankfurt, but he eventually left that organization to become the scientific and technical adviser for the journal *Metalwirtschaft* (*Metallurgical Economy*), headquartered in Berlin. In this position, he was able to travel widely throughout Europe, meeting many members of the scientific and engineering communities, including Peter Kapitsa and others in the Soviet Union.

In 1932, he joined the staff of the world-famous Springer-Verlag publishing house and expanded his activities and travel. Many of the

[23]Arnold Kramish, *The Griffin* (Houghton and Mifflin: Boston, MA, 1986). I am grateful to the author for information concerning his discussions with Riehl in the mid-1980s, during the preparation of his book. This includes the privilege of reading a letter from Riehl to Kramish in which Riehl discussed his friendship with Rosbaud and matters such as the spectrum of attitudes about the Nazi government that prevailed in the German population.

Howard P. Robertson, who served in London during World War II as a key link between the United States and the United Kingdom on matters related to scientific intelligence. He stayed on at General Eisenhower's headquarters until 1946. (Courtesy of the Physics Today *collection of the Niels Bohr Library at the American Institute of Physics.)*

friends he acquired in the process were in high places in their own countries. Moreover, he soon came to know individuals in Germany who were both in and out of the Nazi party but who, like he, had come to despise the regime.

Rosbaud had taken a strong dislike to Hitler and his movement long before Hitler came to power, although he kept his deep detestation well hidden. His wife was Jewish, but that was only one source for his concern and abhorrence. He saw clearly that Hitler was ethically unprincipled, had no appreciation of the rich cultural and intellectual traditions of Europe, and had over-arching military ambitions that could lead to enormous, brutal bloodshed and the destruction of much that Rosbaud valued in the European world. He was one of the all too few individuals who took seriously from the start the programs that Hitler envisioned in *Mein Kampf*.[24]

One of Rosbaud's first activities after the Nazis came to power was to aid the departure of those in the Jewish community who either desired or were forced to leave. While in this relatively dangerous service, he began working closely with sympathetic members of the British embassy in Berlin, particularly those linked to British intelligence organizations. Along the way, he made certain that his wife and children were sent to safe havens in England. As World War II approached and finally exploded, his clandestine activities expanded almost without limit and became more and more the dominant part of his life. It is a miracle that he survived the war.

His role as publisher's agent gave him a cover for espionage as ideal as one could hope for because it permitted him to travel to neutral countries such as Sweden and Switzerland and to those countries that had the misfortune to fall under German domination. In the process, he developed a remarkable network of informers, both spies and others, and a set of highly secret communications links, almost all of which eventually led to intelligence agencies in the British Isles. Frequently the intrinsically valuable information he relayed was not appreciated in England at the time it was received but proved to be vital later on. For example, he accumulated and passed on information concerning developments of aircraft and rockets that were underway at Peenemunde long before the British were able to grasp the significance of what he was telling them. He revealed the ghastly nature of activities in the worst concentration camps soon after they were created, taking his life in his hands to serve as a first-hand witness. He aroused the suspicions of the Gestapo on several occasions but managed to squirm out of their clutches with the help of high-placed friends. On at least one occasion, he vanished from Berlin for a period to allow suspicions to cool off. He often traveled to occupied countries conspicuously wearing a party uniform, as though he were a high-placed official and member of the party. He undoubtedly incurred the hidden wrath of uninformed citizens in the process, but his contacts understood his reasons.

Rosbaud had excellent relations with almost all the members of the Uranium Club and followed the work of the club closely. He had helped Otto Hahn spirit Lise Meitner out of Germany to Holland in 1938. He became an intimate friend of Walter Gerlach, who eventually

---

[24]Another such individual was John von Neumann, who is described in Norman Macrae's *John von Neumann* (Pantheon: New York, 1992).

headed the civilian agency that took over responsibility for the uranium programs. Gerlach frequently sought his advice on many major issues. Werner Heisenberg was one of the notable exceptions among members of the club. He was suspicious of Rosbaud, believing perhaps that his basic links were with the Gestapo, with whom, as mentioned, Heisenberg had occasional difficulties.

It was not Rosbaud's fault that the Western Allies failed to appreciate the fact that the German experimental nuclear fission program was essentially at a standstill by the end of 1942. The politically oriented leaders in the United Kingdom and the United States were inclined to distrust information that was not secured by on-site inspection made by individuals in whom they had well-justified confidence. In particular, General Leslie Groves decided with support from the White House that he would be convinced about the status of German developments only when his own specially selected and trusted staff had made personal inspections. Too much was at stake to rely on a complex foreign espionage network. Although Allen Dulles's intelligence agency, linked to the U.S. embassy in Switzerland, might have been of help to Groves, its cloak-and-dagger mindset was highly oriented toward political, military, and economic affairs.

One great unknown is the extent to which Rosbaud had a branch of his network leading to the Soviet Union as well as to England. We probably will never know, and the matter is undoubtedly irrelevant because the pro-Soviet spies in the British Foreign Office—such as Guy Burgess, Donald MacLean, and Kim Philby—were almost certain to keep the Soviets informed of much important information gained from Germany, including that related to uranium research. Thus, Rosbaud, with his close links to the Uranium Club, may well have been responsible indirectly, if not directly, for the seizure of Nikolaus Riehl by Lavrenty Beria's organization almost as soon as the Soviet army stormed into Berlin. We do know that Beria's organization tried to kidnap Rosbaud at the end of the war in Europe, but he was rescued by an alert Alsos military team.

Rosbaud laid as low as possible immediately after the war and was finally brought to England under cover by military intelligence at the end of 1945. There he picked up his activities as a book publisher, even working for a brief period with the notorious Robert Maxwell, with whom he was not congenial.[25] He died in 1963 and, by his wishes, was buried at sea.

Samuel Goudsmit retained a deep affection and interest in Rosbaud for the remainder of his career and arranged for him to be the

---

[25]Robert W. Cahn gives an account of Rosbaud's professional experiences while in England in *European Review* **1994**, *2(1)*, 34.

first recipient of the John T. Tate Medal of the American Institute of Physics in 1961 at a special ceremony. The archival files of the Bohr Library of the American Institute of Physics contain copies of informative correspondence between Rosbaud and Goudsmit.

In his book, Kramish states, on the basis of an unrevealed source, that a very ugly relationship developed between Heisenberg and his mentor Max Born after 1933. This seemed so unlikely to me that I investigated the matter further, as described in Appendix B to this introduction.

not true

Goudsmit and Alsos

The speed with which the Alsos team sought out and found its major quarries was, to no small extent, the result of the prior knowledge and experience of its chief scientific member, Samuel Goudsmit, who knew with considerable precision what and whom to look for in Europe. He thereby spared the group from what would otherwise have been much wasted time and effort in random searching. The team was also aided by the fact that most of the individuals and equipment they were seeking had been evacuated from the vicinity of Berlin to western Germany, which fell under British, French, and American control. In contrast, circumstances prevented the team from entering the area around Berlin until July 1945, more than two months after the cessation of hostilities in Europe. By that time, the Soviets, with their own foreknowledge, had been able to examine and remove anything related to the uranium program—including the key individuals linked to the program at the Auer Company.

Actually, Pash made several forays to the vicinity of Berlin well before the Alsos team as a whole entered the metropolis. He was searching for the French radium standard that had been purloined by the Germans. Although he found a specimen of radium in what ended up as the East Zone of Germany, it is not clear that it actually was the French standard (see footnote 22 on page 45). He enjoyed fraternizing with the Russian officers during these journeys; he was of Russian descent and fluent in their language.

Goudsmit's background and personal experiences as a member of Alsos deserve mention here because he was well prepared for his role and entered into it with much zeal for personal reasons as well as to make a contribution to the national effort.

Harrison M. Randall was head of the department of physics at the University of Michigan for 23 years, starting in 1917. Although personally a modest man, he had great ambitions for the department and sought to acquire as many outstanding young scientists as opportunities and funds made possible. Through good fortune, he succeeded in convincing two brilliant Dutch physicists, Samuel Goudsmit and George Uhlenbeck, both in their mid-twenties, to join the staff. They may have thought that their sojourn to the United States would be temporary while they waited for more attractive opportunities to open up in Europe, but fate determined that they would serve out their careers in the United States, except for a brief interlude when Uhlenbeck held an appointment in Holland.

Goudsmit and Uhlenbeck were highly celebrated in their professional community by the time they came to the United States because they had had the temerity to propose, in a path-breaking publication, that the electron had an intrinsic twofold degree of freedom, which they associated with "spin," as if the electron were a spinning top and could rotate in either one of two directions.

One would have expected the two to share a Nobel Prize for this major discovery, but they were never so honored. Tradition has it that another young Dutch physicist, R. de L. Kronig, had earlier drawn the same conclusion independently but had been persuaded not to publish a paper on the topic by the highly esteemed and influential physicist, Wolfgang Pauli. The problem that bothered Pauli on that occasion was eventually resolved with the development of a new relativistic version of wave mechanics by Paul Dirac. Legend has it that Pauli wrote to the Nobel award committee stating that it would be unfair to Kronig to give an award to Goudsmit and Uhlenbeck under the circumstances, even though the discovery was made independently.

Goudsmit soon became a central figure in the rapid development of science, particularly physics and chemistry, in the United States in the period preceding World War II. Especially significant were the summer seminars in theoretical physics and chemistry at the University of Michigan. Many of the notables from around the world (see photograph on page 37) were invited for congenial, productive gatherings at which a wide range of topics was explored in depth. Once World War II started, Goudsmit took a leave of absence to join the Radiation Laboratory at the Massachusetts Institute of Technology, where he soon became a major contributor as well as ombudsman in a dynamically creative institution devoted to the development of radar.

Before the Normandy landing, Groves had sent an intelligence team headed by Boris Pash into Italy with the hope of obtaining information from Italian scientists about German developments in the field of nuclear energy. Pash was accompanied by the head of intelligence in Groves's office, Robert Furman. Although the top scientists in Italy such as Edoardo Amaldi and Giancarlo Wick were pleased to cooperate, the results were disappointing, in part because of the inevitable gulf in style of communication between the typical military officer and the civilian scientist. It was concluded that a high-ranking American scientist, intimately familiar with Europe and its scientific community, should be part of the team that accompanied the forces that invaded northern Europe. After much consultation, involving Groves and scientists in the Executive Offices of the White House, Goudsmit was selected to be the chief scientist because he had a broad knowledge of modern physics and of physicists on an international scale and a detailed familiarity with Europe and its major languages. Moreover, he did not have intimate knowledge of the status of the accomplishments of the Manhattan District that could be divulged if he were captured.

Edoardo Amaldi who, with Giancarlo Wick, was interrogated by Boris Pash and his military colleagues concerning the German nuclear program as the Alsos team followed the advance of the Allied armies into Italy. Amaldi became a leading statesman of European as well as Italian science in the postwar era. Had World War II not occurred, it is possible that controlled nuclear power would have been developed first in Europe through the cooperative action of an international European group in which Amaldi would have played an important role. (Courtesy of the Marshak Collection of the Emilio Segrè Visual Archives of the American Institute of Physics.)

Having spent most of the war on problems related to radar, Goudsmit was undoubtedly pleased to accept the special assignment offered by the Alsos mission because it represented an exciting new adventure. Moreover, he had, through intelligence and personality, all of the attributes necessary to become a diligent sleuth. There is little question, however, that the psychological stresses he experienced while carrying out the mission were enormous. He was returning to the Europe in which he had grown up and achieved his early successes. It was now shattered and in enormous disarray, with evidence on all sides of the worst horrors human beings can inflict on one another. Moreover, he had no solid clues concerning the fate of his parents. Meanwhile, the information regarding the enormous scale of the Holocaust was trickling in. He did know that his parents had been taken from their home in Holland to a concentration camp in 1943 and now feared for the worst. A visit from Paris to Holland, once that country had been freed, confirmed his fears. Some of his Dutch friends had appealed to Heisenberg because of his stature in German science to see whether he could help rescue Goudsmit's parents, but whatever was done proved to be too little too late.

His emotions finally overwhelmed him in Strasbourg when, on top of what he had discovered in Holland, the Alsos team encountered direct evidence that a group of German physicians assigned to the medical school in Strasbourg had carried on experimentation on human captives—treating them much as laboratory animals—in complete, brutal violation of the ethics of the medical profession. According to Furman, Goudsmit became so distraught over the combination of inhuman practices that were unfolding and the knowledge of the fate of his parents that he needed a respite in the United States to recover equilibrium before going further. Little wonder! Knowing almost firsthand of the anguish Goudsmit had suffered while playing out his role on the Alsos mission, I never felt free to probe the matter with him in any detailed way. As was the case for many scientists and others from the occupied countries, such as Holland, Denmark, and Norway ,that had tried to remain neutral, what had once been a reasonably high level of admiration for Germany and German culture was transformed to repugnant repudiation.

Cassidy (footnote 5 on page 3) notes that Goudsmit, in his later years, was willing to recognize that most scientists as citizens will feel an obligation to work on behalf of their country. He was, however, unwilling to forgive Heisenberg for not trying to take more effective steps on behalf of his parents.

In judging Heisenberg's sense of loyalty to his country, family, and colleagues, in spite of Hitler's style of leadership, one may recall that relatively few of the southern military officers who had served in the U.S. military, including those who had received their training at West Point and

Annapolis, chose to serve with Union forces in support of the elevating concept of a United States of America once the Civil War had begun.[26] They preferred to support "home and hearth." Deeply entrained sentiments prevailed. The decision made by Robert E. Lee was typical.

By the time the war in Europe ended, the major mission assigned to the Alsos team was essentially completed, with the exceptions of bringing together the select group of German scientists who would end up at Farm Hall and conducting interviews with other German scientists as circumstances permitted. The second part of this task extended well into the following months. In their expeditions through Germany after the end of the war, Goudsmit and Pash found it convenient to use the offices of the Field Intelligence Agency Technical in Frankfurt as one of their bases (see footnote 22). As a result, Robertson and I, who were stationed in that office, not only saw them frequently but made several expeditions with them to sites of special interest to all of us.

The Alsos team actually did not enter Berlin until the end of July. By that time the Soviet occupying group had removed anything in their zone they thought to have value—including Riehl and his colleagues. Goudsmit reports that he visited the "chief chemist" of the Auer Company but learned nothing new. That individual would presumably not have been a prominent member of the Uranium Club.

Although Goudsmit retained links with Groves and the Manhattan District until the demise of that agency, which was replaced by the Atomic Energy Commission at the beginning of 1947, he felt that as a citizen and scientist he had an urgent need to inform the public about issues that had come to light as a result of the work of Alsos. The result is a somewhat restrictively, but emotionally, written book, *Alsos* (see footnote 17 on page 34), apparently published against the wishes of Groves. Perhaps to "protect" Paul Rosbaud, Goudsmit does not discuss him, except in reference to a conversation Rosbaud had with Walter Gerlach near the end of the war. Moreover, Furman is referred to only as "the mysterious major," presumably because Furman had insisted that he not be made a part of the book, in keeping with his or Groves concept of protocol. Actually, Groves bore no long-lasting grudge. In

---

[26]Two (among several) highly notable exceptions were General George H. Thomas, a Virginian, known as the "Rock of Chickamauga" and the successful commander of the Union Army in the critical battle of Nashville, wherein the Confederate Army under General John B. Hood, attempting an invasion of the North with the hope of loosening Grant's grip on Richmond, was defeated. The other officer was Admiral David G. Farragut who brilliantly headed the Union Navy. Actually, Farragut was born near Knoxville, Tennessee, which was a center of activity for anti-secessionists early in the war. This is not to say that Heisenberg possessed the qualities of stately gallantry we associate, for example, with General Robert E. Lee.

The dismantling of the Heisenberg–Wirtz experimental reactor at Haiger-loch by the Alsos team in 1945. (Courtesy of the Goudsmit Collection of the Emilio Segrè Visual Archives of the American Institute of Physics.)

his own book, *Now It Can Be Told,*[27] he does not mention any regret he may have had originally for Goudsmit's action. Moreover, he always enjoyed the occasional reunions of the Alsos group, which continued into the 1960s and which I attended.

Part of Goudsmit's purpose in insisting that he proceed with the book was to emphasize aspects of the inhuman behavior of the core of national leaders in Germany that he had witnessed personally—behavior that, among other things, had cost him his parents under what undoubtedly were unthinkably cruel conditions. Beyond this, however, he wished to stress the vast difference in the way in which the scientific community was employed in Germany and in the United States to the disadvantage of the former and the advantage of the latter.

In retrospect, this difference was not only a result of government actions in the United States—although great intelligence and compe-tence was employed by the leadership in the White House—but a sub-

---

[27]*Now It Can Be Told* (Harper and Brothers: New York, 1962).

The Alsos team excavating the hidden store of cubicle uranium fuel elements for the Haigerloch reactor. Samuel Goudsmit is seated in the pit. (From the Goudsmit Collection of the Emilio Segrè Visual Archives of the American Institute of Physics.)

stantial part rested on the sense of unity that pervaded the American scientific community at the time. The bombing of Pearl Harbor and the simultaneous declaration of war on the United States by both Japan and Germany helped produce that unity, but fear of the anti-intellectual, ruthlessly demagogic world that would result if the Axis powers were victorious played an even more fundamental role. This concern was reinforced by the arrival of many refugee scientists from Germany and Italy, who not only fit in remarkably well with the rapidly evolving scientific structure in the United States but who had been regarded as "superfluous" by the leaders in their own homelands. Most scientists in the United States not only felt a sense of obligation to participate in military research but found effective mechanisms for doing so readily at hand. The combination was the key to the successes that followed.

I saw a great deal of Goudsmit following the war. We not only worked in close collaboration through the American Institute of Physics and the American Physical Society, but Goudsmit was a visiting professor at what is now The Rockefeller University. I served the university first as a member of the board of trustees starting in 1964 and then as president for 10 years until 1978, the year of Goudsmit's death. Goudsmit's former colleague, George Uhlenbeck, joined the Rockefeller University as a resident professor in 1960. One of the most moving moments in my career at the university occurred in 1978 one or two days after Goudsmit's death had occurred and been announced in the newspapers. I chanced to meet Uhlenbeck while crossing the campus; after exchanging a few words of greeting I said, "We are all sorry to hear about Sam." Uhlenbeck queried, "What about him?" I responded, "Oh, I am sorry, you have not heard. He died two days ago." Uhlenbeck reeled about and almost lost his balance.

Although there was considerable fear in top government circles in the United States that the Germans might develop an effective nuclear weapon during World War II, it was also widely believed in the same circles in 1945 that the Soviets would require 20 years or so to match what had been achieved in the United States during the war. Because the Soviets knew that a bomb could be constructed and had information released in the Smyth Report—which was published immediately after the end of the war and provided much detailed technical information about the program of the Manhattan District—Hans Bethe[28] and I offered the contrary opinion that the Soviets would achieve that goal in about five years. Actually they required only four. At the time we wrote our paper, Bethe and I had no knowledge of the level of effectiveness of Soviet espionage.

---

[28]Our paper has been reprinted in Hans A. Bethe's *The Road from Los Alamos* (American Institute of Physics: New York, 1991).

Very little overt criticism has ever been raised by scientists outside the Soviet Union regarding the willingness of so many Soviet scientists to provide their leaders with a full arsenal of nuclear weapons. The moral issues were finally raised, at great personal risk, by heroic participants in those programs such as Andrei Sakharov,[29] once parity had been reached.

One might wonder why Riehl waited so long to have his account of his years as a Soviet captive offered for publication. There are probably two principal reasons: He was busy with research and lecturing at the Technical University of Munich until his retirement in about 1970, by which time he had prepared a more or less final draft for private reading. But perhaps more important, he did not want to do anything that might offend Soviet friends whom he admired so much. The changes in the Soviet Union with Gorbachev's leadership and the advent of Glasnost greatly altered the situation.

It is sometimes said that one important measure of the quality of an individual is the extent to which he or she can preserve a sense of "inner dignity" under very trying circumstances. Nikolaus Riehl scores very high when so measured.

Appendix A: Use of the Bombs on Japan by the United States

During the days I spent in the office of Secretary H. F. Stimson in the spring of 1945, while preparing to join H. P. Robertson at General Eisenhower's headquarters, first in Versailles and then in Frankfurt, I devoted considerable time to discussions with the staff regarding its views of what our forces would face in the forthcoming invasion of Japan. The possible use of nuclear bombs obviously was not included. Up to that time, some 15% of our military casualties had occurred in the invasion of Okinawa as a result of the absolutely fanatical defense put up by the Japanese forces and civilians in the protection of this outlying island. Japanese military casualties, not to mention civilian ones, averaged seven times our own in such engagements.

Estimates were made in the Pentagon at that time that in a very probable "worst case" the invasion of Japan proper would last at least a year and involve a million U.S. casualties. Anyone who doubts the potential for fanaticism of the Japanese militarists at that time should read Soichi Oya's *Japan's Longest Day* (Kodansha International:

---

[29]A review of Sakharov's turbulent life and works appears in *Sakharov Remembered: A Tribute by Friends and Colleagues*, edited by Sidney D. Drell and Sergei P. Kapitza (American Institute of Physics: New York, 1991).

Andrei Sakharov in the late 1980s after he had gained freedom from house arrest at Gorki and was allowed to travel to foreign countries. He is shown signing the membership book of the National Academy of Sciences in Washington, DC. (Courtesy of the National Academy of Sciences.)

Tokyo, Japan, 1968). This book describes the attempt of the fanatics after the nuclear bombings to prevent the emperor from reaching a Tokyo radio station in order to request his forces to lay down their arms. Arguably, the dropping of the first bomb, more or less in the manner in which it was employed, was inevitable and necessary under the circumstances—for the long-run sake of the Japanese people as well as our own. One might, however, have waited somewhat longer before using the second bomb in order to give the governmental machinery in Japan time to operate. In any event, President Truman, who was anything but a brutal man, made the best of a difficult situation as he saw it.

Nikolaus Riehl's succinct, thoughtful comments on such moralistic matters are contained in Chapter 14 of his book. He separates the responsibilities of scientists from those of the politician, although he is well aware of the failings on both sides. [A clear picture of the Japanese military view "fight with honor to the last person" is given in the book by Colonel Hiromichi Yahara, *The Battle of Okinawa* (John Wiley & Sons: New York, 1995).]

Appendix B: Heisenberg and Born: Questioning Kramish's Sources

Kramish's *The Griffin* (see footnote 23 on page 47), although very valuable for its broad sweep and much of its detail, is somewhat quirky in the sense that one is occasionally left to question the reliability of his sources of information. Moreover, some facets of the history, or of the characteristics, of the individuals appearing in the book may seem distorted out of what could be a personal, emotional bias on the part of those sources. One example, which I regard to be of the utmost importance, is the following:

> By 1934, Born had settled in Cambridge, and Heisenberg came to bring an official message from the Nazi government, the first of many official missions Heisenberg would carry out for the Nazis. He had obtained permission for Born to return to Göttingen. As Born recorded the incident in his memoirs, "I would not be allowed to teach but could do research. I was rather bewildered by this sudden offer, which to a homesick immigrant sounded quite attractive. After a few moments of consideration I asked, 'and this offer would include my wife and children?' Then Heisenberg became embarrassed and answered: 'No, I think your family is not included in the invitation.' This made me very angry."

> Subsequently, Born did go back in an unsuccessful attempt to retrieve some of his possessions. One of his associates had vividly recalled Born's brief return to Göttingen: H. was by then a professor in Göttingen, and when the Borns went to visit him, they were met with anti-Jewish sneers and obscenities, and in the end H. spat on the floor at Max Born's feet! . . . When the Borns returned (to England) and . . . I warmly inquired about their trip and how they had fared, very reluctantly Max Born confided in me about this great shock . . . I was both horrified and profoundly angered. . . . Later Mrs. Born gave me her version and ended with a statement that I have never forgotten. She said simply at the end, "And my poor Max wept."

This last paragraph is offered "anonymously"; however, the charge made with respect to Heisenberg's behavior is so important that it, at the very least, deserves being more fully documented and not left to hearsay. Moreover, the source should be made known in the case of such a serious personal accusation. Heisenberg may be regarded as a misguided or overzealous patriot or nationalist—he world is full of them—but we do know that he had great admiration for Max Born (photograph on page 23). Offering anti-Jewish sneers and obscenities and spitting at Born's feet are acts that do not appear to fit in with the true relationship between the two scientists that we might expect.

In fact, Heisenberg held a professorship at the University of Leipzig and not at Göttingen at the time of the incident described above. Moreover, in a letter to his old friend Otto Westphal dated September 2, 1994, Max Born's son, Gustav V. R. Born, who is now director of the William Harvey Research Institute of Saint Bartholomew's Hospital Medical College in London, has the following to say with regard to the paragraphs from Kramish's book:

> Professor Otto Westphal
> Chemin de Ballallaz 18
> CH-1820
> Switzerland
>
> Dear Otto,
>
> Many thanks for your two letters of 9th and 30th August. Needless to say, what you write is of the greatest interest to me. I have for years now "broken my head" (as they say in Germany) about the relationship between my father Max and Heisenberg. They were involved with one another, after all, for so many years: from 1923 when Heisenberg came to Göttingen as Max Born's Assistant right up to my father's death in 1970. I am sure that the relationship changed enormously over the years, as it was bound to do because of all the many things that happened.
>
> I myself have quite clear memories of Heisenberg, both from my childhood in Göttingen and through his visit to us in Cambridge—referred to in my father's autobiography which you now quote—to the post-War period when Heisenberg and Planck came to visit us in Edinburgh, shortly after my return from War service and shortly before Planck died. I think I could say quite a lot about all this but, like you, I have never yet found time to write about it. Again like you, I hope to be able to do that before it is too late. Right now I am still deeply involved in direct research (happily successfully!)
>
> To the specific point from the book by Kramish (which, incidentally, I have not read and should much like to read), I have been told about this particular episode before. The first

part is evidently true, because it comes from my father's autobiography. However, the subsequent part I have serious doubts about. To the best of my knowledge and recollection, my father never visited Göttingen again, after we left Germany in 1933. I know for certain that my mother Hedi and I did return—I believe it was in 1936—to go through and destroy papers remaining in the loft of our house, Wilhelm-Weber Strasse 42. It became quite a horrific visit because we decided to spend a few days' holiday in nearby Höxter and walking in the Solling Forest; and that area had by then been profoundly Nazified. It was quite frightening, and the relief in returning to Britain was tremendous.

From my entire recollections I cannot conceive that Heisenberg would have produced anti-Jewish sneers and obscenities and would have spat on the floor in front of his former Professor, whom to the best of my knowledge he revered.

Enough of this just now because of pressure on my time; but I wanted to let you have a worthwhile answer as quickly as possible.

With many thanks again for writing so interestingly and wish best wishes to you and your,

Yours as always,
Gustav

This letter was written in response to one Otto Westphal had written to Gustav Born enclosing a copy of the foregoing paragraphs from Kramish's book which I had provided Westphal. It supports the view that the incident described in Kramish's book could not have involved Heisenberg. Additional confirmation of this point of view has been obtained through discussions I have had with members of the Niels Bohr Archive in Copenhagen who have long, intimate familiarity with the relationship between Born and Heisenberg.[30]

---

[30]See *also* Abraham Pais's *Niels Bohr's Times* (Oxford Press: New York, 1991) for additional material on the relationship between Bohr and Heisenberg.

TEN YEARS IN A GOLDEN CAGE

Chapter 1
Introduction[1]

In this book, the reader will find my recollections of the years 1945–1955 when I was retained as a captive of war and leader of a group of German scientists and technical specialists in the Soviet Union. Thanks to the level of my activity in that country and my complete command of the Russian language, I came into closer contact with Soviet life than normally would have been possible for a foreigner. Indeed, this was true in several areas: at the level of the government, the management of ministries, the research institutes and factories and, not least, the daily lives of the Soviet peoples.

I was frequently asked by many individuals to put on paper my experiences and the situations I had witnessed, because it appeared that they could have political and historical interest. In following this advice, I do so without any desire to be either extensive or exhaustive concerning my Soviet relationships at the time of internment. Instead, I plan to dwell on the personal experiences and impressions of an unbiased observer who desires to put matters into some degree of order and analyze them, granting that the bare facts often continue to astonish even me. What is related here is, indeed, history; however, I have chosen to present it in the form of anecdotes.

I have adopted what may be regarded as an ironically amusing style of presentation in many places in this book, disregarding the aca-

---

[1] It is recommended that the previous introductory material be read before starting Riehl's account, because it gives the background of Riehl's professional career before 1945.

demic traditions of formality.[2] Actually, the period in which we had to deal with Stalin's government was anything but amusing. It is, however, well-known that dictatorships often place one in ridiculous situations and that on such occasions one exhibits a combination of frustrated aggravation and amusement. In focusing on my personal experiences, I am avoiding a descent into what might be regarded as the production of an autobiography. I will also do my best not to add to the many literary accounts of lives that are decorated with youthful pictures or colorful representations of the individual involved. Such creations rightfully belong to artists, athletes, or politicians. Instead, I hope to provide a better understanding of what was achieved in the course of my life in Russia for both the uninitiated and the experts. In keeping with this spirit, I will offer no more than the following pithy account of my professional career prior to my period of residence in Russia that started in 1945.

I was born in St. Petersburg, Russia, in 1901 as the son of a professional engineer employed by the Siemens and Halske Company who was in charge of operations there. I was resettled in Berlin with my parents who were classified as German citizens at the time of the Brest–Litovsk treaty of 1918, which ended the conflict between Germany and Russia in World War I. Although I attended the German language schools in St. Petersburg, I was completely at home in the Russian language. After completing work for a doctoral degree at the Institute of Otto Hahn and Lise Meitner in Dahlem in 1927, I accepted a position at the Auer-Gesellschaft, a very distinguished firm in Berlin named after the famous Austrian inventor, Auer von Welsbach, and initially devoted to the exploitation of his inventions.[3] At that time the principal activities of the company were centered on the development and manufacture of the Welsbach mantles containing thorium and cerium oxide, which were used to enhance the illumination of gaslights, and on the manufacture of gas masks that removed toxic substances from the air. As a result of my initiative, the program was expanded to

---

[2]As part of his psychological armor for survival during the exceedingly difficult period between 1933 when Hitler gained power and his return from the Soviet Union in 1955, Riehl found it necessary to cultivate an ironical or sardonic sense of humor. Because his brand of humor tends to be Russo–Germanic, some American readers have found his "asides" confusing. It is important to realize that Riehl, as a highly rational individual, had to deal in some way with the intense irrationalities he encountered during these very difficult years. His ironical comments, which appear frequently in the book, are to be appreciated in these terms.

[3]The Auer Company ended up in the Soviet-occupied zone of Berlin and became part of Communist East Germany.

include research and development of fluorescent materials and of uranium[4] (I should mention that in the period after World War II the re-es-
tablished Auer Company retained only activities related to the manufacture of gas masks, although that production was expanded in many
directions). I started my work in the division for radioactive substances.
After pursuing the management of various topics in the field of applied
radioactivity, I broadened my area of investigation and initiated the
development of luminescent lamps (often falsely called neon tubes in
popular language). This development, carried out in close cooperation
with the Osram Lamp Company, had major consequences wherein
the Auer Company produced luminescent materials and Osram the
actual lamps. I then turned to other special applications of luminescent
substances, including luminescent dyes, X-ray imaging screens, and
television screens, meanwhile carrying on basic scientific work associated with this entire area of investigation. The conclusions of all of this
work were summarized in my book, *Lumineszenz und ihre Anwendung*
(*Luminescence and Its Applications*), which was translated into several
languages. I succeeded in returning to this beloved area of research
many years later. In 1938 I obtained the special diploma that would
permit me to teach at a university; however, I remained with the Auer
Company and, shortly before the outbreak of World War II, was made
the head of a newly founded scientific division that had the responsibility of finding new areas of research and development and of opening
them to investigation for the benefit of Auer.

Once uranium fission was discovered, it became my responsibility
to take on the problem of developing technology for producing the purest uranium of which we were capable and which might be used for
the release of nuclear energy. This task was particularly appropriate
for the Auer Company because it had considerable experience in similar areas of chemical technology. Adding to our strength was the
anticipated contribution of the Degussa Company, located in Frankfurt, to which the Auer Company belonged at that time. Degussa had
much advanced metallurgical experience and was in a position to provide the final step in the process, namely that of producing metallic
uranium. The Nukem Company, which produced uranium fuel elements for the German market, emerged out of this effort after World
War II. In the meantime, however, I must describe the 10 years of activity related to the production of pure uranium that my colleagues and I
undertook in the Soviet Union.

---

[4]Riehl had previously promoted the production of metallic thorium, which was
used as an addition to tungsten to enhance electron emission from cathodes in
vacuum tubes.

As can be seen, my life alternated in the manner of a pendulum between physics and chemistry, between science and technology, between management and research. As a result, I sometimes refer to myself as a general store clerk. Moreover, my life also resembled a pendulum in the geographical sense. I spent my childhood and youth in St. Petersburg, 25 years in Berlin, 10 years in the Soviet Union, and finally 30 years in Germany.

When I employ conversational speech in quotation marks in the following, as I will do in order to improve readability, it is not to be implied that I am providing accuracy in the strict stenographic sense, because the translation of Russian into another language requires considerable restructuring of sentences. I will, however, attempt to adhere rigidly to the intention and mood that accompanied the statements in such cases.

Comment: The greater part of the following text was written before 1970 and remains unaltered. A few details were added later to give emphasis and clarification to the present state of affairs. This has been done, for example, in Chapter 3 in connection with the episodes related to the physicist Peter Kapitsa. Moreover, many additions and expanding explorations were stimulated by comments and questions raised by the large circle of individuals who read the original manuscript. Such changes are particularly valid for Chapter 14. The description of the situation in the Soviet Union given in this book focuses almost exclusively on the period in which Stalin was dictator and does not, for example, reflect the changes associated with the emergence of Gorbachev. I should add that the skeptical judgments regarding the possible liberalization of the Soviet economy in the future, which appear in the final chapter, 18, have been left unaltered in the quiet hope that my skepticism will prove to be unjustified.

Chapter 2

Transportation to the Soviet Union and the Search for an Appropriate Site for the Production of Uranium

Berlin lay in rubble and ashes. Hitler's thousand-year Reich had ended. A portion of my colleagues, as well as my family and I, were housed in numerous villages in the vicinity of Rheinsberg in the Brandenburg Mark. We had brought some of our equipment with us, but work languished.

We did not know which national troops would enter and occupy our region. We learned from British radio that Berlin would be occupied by all four victorious powers. Common sense led one to expect that the four zones of occupation would form wedges abutting one another. In our naive, politically unsophisticated manner, however, we considered it doubtful that the Americans, British, or Russians actually would occupy our local area of Rheinsberg northwest of Berlin. It required the marvelous, long-range viewpoint of the politically sophisticated Western statesmen to decide to create an island of Berlin and, in the process, transform it into a continuous source of conflict between the West and the East. In any event, the Russians actually did come to our area.

In the middle of May of 1945 two colonels of the Peoples' Commissariat of the Interior (NKVD) suddenly appeared from Berlin along with my friend K. G. Zimmer, who was at that time actively linked partly to my scientific organization and partly to the Kaiser-Wilhelm[1] Institute in Berlin-Buch. The colonels requested that I join them for a "few days" of discussion in Berlin. The few days lasted for 10 years.

---

[1]The pre-war name of what are now the Max Planck Institutes.

It soon became evident that the colonels were not really colonels. Actually, they were two physics professors outfitted in colonel's uniforms. One was L. A. Artsimovich, who later became very prominent because of his work in the Soviet program devoted to nuclear fusion. The other was G. N. Flerov, a co-discoverer of spontaneous fission of uranium, that is, fission that is not induced by the capture of a neutron but that occurs naturally. These and other Soviet civilians who were sent to Germany at the time were outfitted in military uniforms so that they could circulate through military channels as their needs required. Many of them appeared quite droll in their uniforms. That was particularly true of the otherwise distinguished physicist, J. B. Khariton, who was wearing much too large a military cap. Fortunately, his ears extended sufficiently to prevent the cap from coming down completely over his slender scholarly head.[2]

Returning to the realities of life, it should be mentioned that K. G. Zimmer was released when we arrived in Berlin; however, I was taken to a guarded home in Berlin-Friedrichshagen and resided there for a week. I was then taken to the headquarters of the Auer Company in Berlin where the disassembly of all removable plant equipment and apparatus was well underway. The staff of Lieutenant General A. P. Zavenyagin, the current Acting Deputy Peoples' Commissioner in the NKVD—that is, the deputy of Lavrenty Beria—was located in Friedrichshagen. All Peoples' Commissariats were later renamed Ministries so that NKVD was transformed to MVD. (Actually, the more complete "genealogy" of the establishment is as follows: Cheka–GPU–NKVD–MVD). Later on Zavenyagin became the Soviet Atomic Minister. We were to encounter one another frequently in the future. His ministry was actually given a quite different title as a matter of camouflage; however, I will continue to refer to it as the Atomic Ministry for the benefit of brevity and clarity.

Perhaps this is the place to add a few additional explanatory words about the NKVD and MVD, because the "genealogy" given in the previous paragraph is incomplete. The MGB and the KGB, which serve as the official Soviet security agencies, corresponding to what was the German Gestapo, are associated with and at the same level as the MVD. A more complete account of this entire complex is given in Solzhenitsyn's well-known book, *The Gulag Archipelago*. This book is often criticized as being overly bitter and hateful. I believe, however, that Solzhenitsyn was not inspired to write the book merely out of bitter feeling for the suffering that he and his people had experienced. Indeed, I can confirm the accuracy of his descriptions and evaluations

---

[2]Translator's note: I was attached to General Eisenhower's headquarters in 1945 in a similar way but was more nattily outfitted.

Lev. A. Artsimovich is second from the right in this postwar photograph taken during a visit from the French scientist Frédéric Joliot to the Soviet Union. From left to right are F. Joliot, I. V. Kurchatov, D. Skobeltzyn, L. A. Artsimovich, and A. Alikhanov. (Courtesy of the Emilio Segrè Visual Archives of the American Institute of Physics.)

Georgi N. Flerov in 1945 at the time he and his colleagues called upon
Nikolaus Riehl, beginning Riehl's 10 years of servitude. (Courtesy of R. Kuznetsova of the Archives of the Kurchatov Memorial Institute in Moscow.)

both from my own experiences and from those of many individuals
mentioned in the book whom I have known. Unfortunately!

The component of the MVD into which we German "specialists"
were immediately assigned[3] can be regarded as part of the function-

---

[3]At this stage of events the German scientists were completely unaware of
developments in nuclear energy that were taking place in the United States,
unlike their Soviet counterparts. The Trinity Test (July 16, 1945) and the use of
the bombs on Hiroshima and Nagasaki (August 6 and 11) took place two and
three months after Riehl was made captive by the Soviet agents. In the book,
Riehl does not relate precisely when he first became aware of the use of
nuclear bombs on Japan, but he undoubtedly knew of it soon after it took place.
For example, he mentions receiving a Russian translation of the Smyth Report
soon after it was published in the United States.

Georgi N. Flerov as a soldier in 1941 at the age of 28. While serving in the army near the battle front in 1942, Flerov noted that U.S. scientific journals no longer contained accounts of research involving nuclear fission. He concluded correctly that most nuclear scientists had been absorbed into secret research dealing with nuclear chain reactions. In frustration over inaction at lower levels among Soviet scientists and officials when he wrote to them about this observation, he eventually addressed a critical letter to Stalin that did receive attention. (Courtesy of R. Kuznetsova of the Archives of the Kurchatov Memorial Institute in Moscow.)

ing of a colossal governmental activity in which the great majority of the participating individuals, exclusive of the guards, were prisoners of all categories, ranging from ordinary murderers to politically suspect university professors. The organization operated in various areas extending from the construction of canals to enterprises concerned with technical developments and scientific research laboratories.

J. B. Khariton and I. V. Kurchatov aboard a ship in 1954. (Courtesy of R. Kuznetsova of the Archives of the Kurchatov Memorial Institute.)

Avram P. Zavenyagin, a deputy of Lavrenty Beria and Riehl's link to the latter. His physical resemblance to Mikhail Gorbachev, mentioned by Riehl, is remarkable. (Courtesy of the U.S. Library of Congress.)

Experts and administrators other than prisoners were obviously a necessary part of such an operation. A Soviet academic metallurgist of technical university rank once said to me by way of explanation, "As you can understand, we have many social rejects in our land. We use such individuals for the development of the country." With the use of the term "rejects" (in the sense of waste from technical production, which in Russian is *otchody*), he described the imprisoned individuals very elegantly. One never learns—.

I was taken to our factory in Oranienburg (north of Berlin), where we had the actual plant for producing our purest uranium oxide, on two occasions before being taken to the Soviet Union. (High purity is

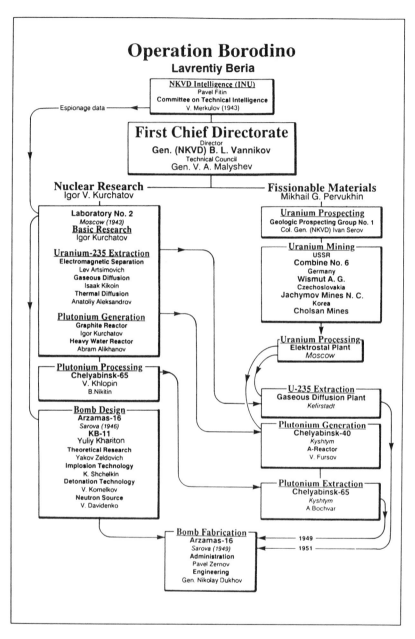

The organization chart of Operation Borodino, the nuclear bomb program within Lavrenty Beria's ministry. (Reproduced with permission from Target America, by Stephen J. Zaloga © 1993. Published by Presidio Press, 505 B San Marin Drive, Novato, CA 94945.)

the primary requirement for uranium that is to be used in a nuclear reactor.) The works had been bombed twice by American forces and were damaged almost completely. The bombing had taken place just before the end of the war and made no sense at all to us at the time. I felt a definite reaction from my Russian colleagues, however, as we went about the damaged plant. Their response, which involved the shaking of heads and casting of meaningful side glances, was unintelligible to me at the time. I appreciated the reasons for the reactions much later. The Americans had occupied southwest Germany for a considerable time before Berlin was actually taken over. During that period they learned from the group of German physicists held at Haigerloch, including Hahn and Heisenberg, that uranium intended for nuclear reactors was being manufactured at Oranienburg. Knowledge of that source would normally have been of no immediate pressing concern to the Allies, because they already knew that the Germans were far from producing a nuclear bomb. Relationships between the Western Allies and the Soviet Union had already begun to deteriorate during this period, however, making it understandable that the Westerners would prefer not to have the Oranienburg works fall into the hands of the Russians intact. The Russians who accompanied me appreciated the fact that the bombings were directed not against the Germans but against them.

The demounting and loading of everything that was not nailed down or riveted proceeded at full speed. One day a colonel, who was a platinum specialist and who we Germans referred to as the "Platinum Colonel" thereafter, came to me and asked why we were holding back information regarding the laboratory where we had analytical spectroscopic and mineralogical equipment. I stated that we had no special need for such equipment in our work. He replied, "I do not believe that. You must have had such equipment." Somewhat enraged, I replied, "If you do not believe me, we have nothing more to discuss!" The next morning I was approached at a corner by a very neatly dressed young NKVD lieutenant who clearly was a guard, that is, a "professional." He said he had observed the incident and that during the previous evening Zavenyagin had charged the colonel with treating me very roughly. The colonel received a sharp reprimand but explained that he had not intended to be rough, he merely possessed a rough voice.

In this connection, I must mention that it was notable that the "professionals," the functionaries of the security organization, were always especially friendly to me. They gave me advice and provided me with chocolate, tobacco, and other glorious things. As we were about to drive off for the flight to Moscow, an intrepid, burly NKVD lieutenant ran behind the auto, pressed my hand, wished me the best, and shouted the prophetic words, "You will eventually tour about Moscow in your own automobile." I recall that in the period immediately after the October Revolution in 1917, when I was in St. Petersburg, I got along very well with the Communist Secret Police there and even fairly well later on with the Gestapo. I do not know what my popularity with the representatives of this special profession is based upon. I do know that one should approach such individuals on a purely human level, that is, on a plane on which they are biologically and not professionally programmed. One should not display anxiety, appeal to them with juristic arguments, or attempt to curry favor. Moreover, it is occasionally necessary to maintain a firm position. My experiences, however, may be based on the fact that I am what might be called "different," arousing an interest in the "exotic" in the professional security police and that interest and good will are awakened on that account. Who knows whether or not a well-fed tiger would be inclined occasionally to toss a human a piece of meat in good will if it happened to stroll by an individual sitting in a cage.

As early as June 9, 1945, we, that is, a segment of my co-workers and I along with our families, were flown to Moscow. Temporary accommodations were provided, first in a sanitorium near Moscow and then in the Villa "Ozyora," which had once belonged to a Moscow millionaire named Riabushinski. It had been occupied in the 1930s by Yagoda, the head of the NKVD who was liquidated in 1938. The Russians, however, continued to call it "Yagoda's Dacha." The unfortunate Field Marshal Paulus[4] and his staff officers had occupied the same house prior to us following their surrender at Stalingrad. A colossal map with pins showing the front of the battle still hung in the dining room. There was a small shell-battered German battle tank in the park of the villa, which bordered Minsk Boulevard. The tank was alleged to be the one that had approached nearest to Moscow during the war.

In addition to mine, two other groups of German "specialists" were brought to the Soviet Union to work on the problems of nuclear energy. One was led by the famous physicist Gustav Hertz, a nephew of the distinguished Heinrich Hertz, discoverer of electromagnetic waves; the other was led by the well-known electronics engineer Manfred von

---

[4]Field Marshall Paulus surrendered in the Battle of Stalingrad, the turning point of the war on the Eastern Front. Soon thereafter he denounced Hitler's government, gaining much international publicity. This battle, fought with great ferocity, was undoubtedly the most costly in human lives in the history of modern warfare. See William Craig's *Enemy at the Gates—The Battle for Stalingrad* (Reader's Digest–E. P. Dutton Company: New York, 1973).

Friedrich von Paulus, the general who commanded the army that advanced on Stalingrad during the winter of 1942–1943. His actions were directed by orders from Hitler who, in his arrogance, failed to realize the true situation, which would have required a retreat of Paulus's army to a more secure position. Paulus surrendered the army after it had been surrounded and badly defeated in the city in January 1943. He bitterly denounced Hitler for gross mismanagement of military affairs. He remained in the Soviet Union until 1953 when he returned to communist East Germany as a private citizen. In the words of William Craig (see footnote 4 on page 80), " . . . the battle was the greatest military bloodbath in recorded history." (Courtesy of Reader's Digest–E.P. Dutton Company.)

Ardenne.[5] Several individuals were added to these groups, in particular the distinguished physical chemist Max Volmer, the Leipzig nuclear physicist R. Döpel, and somewhat later, P. Thiessen, the ongoing director of the Kaiser-Wilhelm Institute for Physical Chemistry in Dahlem.

A few days after our arrival, while we were still in the neighborhood of Moscow, all of us, that is, Hertz, Volmer, von Ardenne, and I and our wives were invited to the Bolshoi Theater for a performance of Borodin's *Prince Igor*. The stimulating, roaring sound of a victory celebration permeated the theater. Both in the parquet and surrounding us in the various rows were uniformed officers and other representatives of the Western Allied nations, as well as delegations from different ethnic groups in the Soviet Union, all dressed for a celebration. As we listened while standing to the rendition of the Soviet anthem, which was played prior to each performance just after the war, I was seized by a strange mood. The situation seemed unreal. Just a few weeks earlier we were residing in misery in collapsed Germany, and now we were listening to the Soviet anthem in the midst of the victorious clamor of the Allies. Incidentally, we also noted that some of the Western delegates recognized us as Germans during the long intermission to the opera in the foyer and stared at us with interest.[6]

Soon after this event both Hertz and von Ardenne went to the community where they were to work, namely Sukhumi, on the southern Caucasian coast of the Black Sea. My group, however, had still to find a suitable location because we had to satisfy fairly stringent conditions

---

[5]Manfred von Ardenne (see photograph on page 30), a brilliant inventor in the various phases of radio, television, and the electron microscope, has written an extended autobiography in German, *Die Erinnerungen*, published in 1990, as mentioned in the introduction. Among other things this has a section devoted to his years in the Soviet Union. He became engaged in a continuation of the type of research and development he had carried out in Berlin in his private institute, the contents of which were transported to Sukhumi on the Black Sea. He served as special consultant and provided technical support to the Soviet group involved in the development of the fission bomb. He worked independently of but in close collaboration with Gustav Hertz who had his own institute nearby. Unlike Riehl, he returned willingly to Communist East Germany rather than to West Germany in 1955, being aided both by the Communist German government and the Soviet authorities to re-establish his private laboratory in Dresden. Von Ardenne is still very active at the time this translation is prepared. His institute, currently in what constitutes a reunited Germany, is now a center for the special treatment of cancer. As a captive in the Soviet Union he felt strongly that it was appropriate for the Soviet Union to obtain a nuclear arsenal to counterbalance that of Western nations.

[6]Von Ardenne also describes the events of the evening. He concluded that the Western delegates thought of them as German communists who were being befriended by the Soviet officials.

with regard to technical, spacial, and personnel matters if we were to establish a uranium factory, in accordance with our commission. As a result, I traveled about the country during the following weeks, mainly with Zavenyagin and his colleagues, looking for something suitable. Moreover, my co-worker G. Wirths, flew to Krasnoyarsk on Enissei in Siberia, about 10°, or 400 statute miles, east of Novosibirsk, for the same purpose.[7] We continually visited building complexes, mostly abandoned factories. I asked Zavenyagin whether it would not be wiser to build a new building especially designed for our purposes. He said, however, that the country was so impoverished for construction materials, as a result of wartime demands and destruction, that we were compelled to continue to look for an available building.

Incidentally, Zavenyagin wished throughout this process to find us a place in the middle of a beautiful natural setting, while I, with all of my love of nature, pressed for something in the vicinity of civilization, such as a place near Moscow or Leningrad. My concern was based mainly on practical rather than personal grounds because I feared that we would be at a technical disadvantage if we accepted his preferred choice. We were eventually located near Moscow but not because of my influence. The location chosen, as will be seen, worked out very favorably.

On one of the exploratory trips in the upper reaches of the Don, this time unaccompanied by Zavenyagin, I was led to an abandoned vodka factory where convalescing captured German soldiers were lodged. The prisoners operated a furniture factory in which they produced many lovely objects (for example, a complete furniture set for the office of their obviously admired prison commander). In the process they had created a standing floor clock in which the operating mechanism was made entirely of wood. We were warned to avoid the furniture shop because the sawdust on the floor was loaded with fleas. Being zealous in our work and somewhat careless, we proceeded to enter the shop. That evening we returned to Moscow in our special sleeping car belonging to the NKVD. The Russians occupied a large common compartment, whereas I was alone in a single unit. On undressing, I discovered a flea. It was my first experience with this form of vermin. The acquaintance ended with the death of the flea and I slept peacefully. When I entered the common breakfast room on the train the next morning, I described with pride my hunting expedition of

---

[7]According to the video, *Nazis and the Russian Bomb* (see footnote 2 of the introduction of this book), G. Wirths was called back suddenly from the Siberian journey because the Soviet authorities had just learned of the use of the fission bombs at Hiroshima and Nagasaki and were now anxious to accelerate their own plans at any cost.

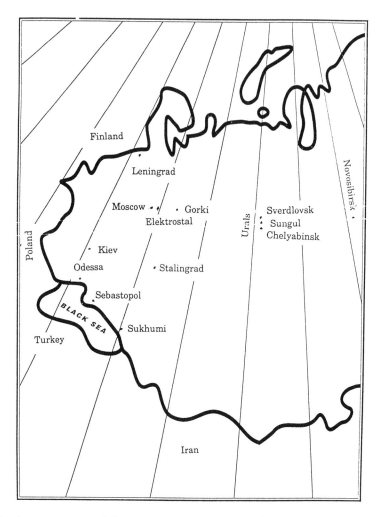

A schematic map of the western portion of the Soviet Union as far as
Novosibirsk. The lines of longitude are separated by 10°, Moscow being
close to 37° east of Greenwich and 56° north latitude. The relative positions of Elektrostal, Sungul (about 55° north latitude), and Sukhumi on the
Black Sea are shown. Sungul is situated in the industrialized area east of
the Ural Mountains which played an important role as a production center in World War II because it was not near the battle zone. Sukhumi is
close to the centers where the research institutes of Hertz and von
Ardenne were located and to which Riehl moved toward the end of his
stay in the Soviet Union. Kyshtym, which experienced a major radiation
disaster in 1957, is close to Sungul. Although the latitude of Sukhumi is
about 44° north, comparable to Halifax, Nova Scotia, the region is semi-
tropical because of shielding provided by the Caucasus Mountains.

The prefabricated wooden homes, manufactured in Finland, in which the German scientists and others lived on the grounds of the Elektrostal works between 1945 and 1950. Riehl's youngest daughter Irene is shown. All photography was subject to strict regulations. Limited photography of the family was allowed, but no pictures showing the factory, the barbed wire enclosure, or the security staff could be taken. (Courtesy of the Riehl family.)

Riehl's eldest daughter, Ingeborg, and her younger sister, Irene, taken at their home in Elektrostal. (Courtesy of the Riehl family.)

the previous evening, but all I received from the group was mocking laughter. They had not slept the entire night and had hunted down innumerable fleas. I returned home with the quiet, proud confidence that the fleas had sufficient sophistication to leave a civilized, middle-European unburdened. My pride, however, vanished at bedtime. To be brief, I had brought about 90 fleas into our home. I shared about 10 of them with my co-workers. It fell to my wife and me to stalk the remainder, both morning and night, during the course of the next week.

We finally selected a respectable munitions factory, which had been decommissioned at the end of the war, as the most appropriate place for our uranium factory. It consisted of many large and small buildings that were distributed in a large marshy woods. It was located at the industrial complex of Elektrostal in the vicinity of the small town of Noginsk (earlier Bogorodsk) about seventy kilometers east of Moscow. The location of this first uranium factory was kept as a strict secret for a very long time. It is no longer a secret, however, except possibly in the minds of a few zealous security guards who retain their attitude out of a sense of duty. The site was selected because, in addition to its buildings, it had a great many skilled workers and much in the way of auxiliary facilities, including workshops, its own power station, and a large parking area for automobiles. The area, unfortunately, was a truly dreadful place in which to reside, as we did for the next five years. The complex contained an electric steel works, after which the site was named, as well as some additional factories that had been moved from the Ukraine.

It is worth noting that we were not the first but the third German group to establish a factory there. Even the munitions factory had been constructed by Germans before the World War I. The single-story stone houses, which had been built for their use, were still preserved. In addition, the electric steel factory had been built by Germans in the 1930s. The director of the munitions factory, a sympathetic general, who remained the director of our uranium factory during the first phase of our construction, recalled the second group of Germans well and related to me on one occasion how one of the alcoholic members of the German group had drunk himself to death. Russia was once again a fateful place for Germans.

Chapter 3

In Elektrostal near Moscow: Initial Difficulties and the Successful End[1]

The production of uranium for nuclear reactors requires, above all, extreme purification of the uranium compound derived from the basic ore (pitchblende) by appropriate chemical means. Very special emphasis is given to removing elements that can capture the neutrons produced in the reactor and stop the chain reaction. The rare earth elements, along with boron and cadmium, must be reduced to an absolute minimum. It is also necessary to remove many other elements. Once that is done, the uranium must be converted, or reduced, to metallic form and then melted and cast into an ingot that has the geometrical form appropriate for the reactor. In some reactor types, it is feasible to use fuel elements consisting of the densest achievable form of uranium oxide. Contamination that could possibly originate from the reducing agents or the crucibles involved in the high-temperature steps must be avoided at all costs.

When we began to devote our attention to these problems in Germany, the technologies had not been worked out because there had been no use for very pure uranium compounds or uranium metal, par-

---

[1]It must be emphasized that Riehl had no more than what might be termed a worm's eye view of the rapid development of the Soviet arms program in the years between 1945 and 1952, when he was most diligently at work for the Soviet Union and before he and a few of his colleagues were allowed to reside in relative luxury on the Black Sea, albeit in a highly restricted way, while waiting to go home. Nevertheless, his keen insight into the part of the Soviet world he was able to witness, when coupled with his command of the language and familiarity with the people, gives his story very special value.

ticularly for extremely pure metal. Uranium-containing ores were pro-
cessed mainly to obtain the radium present. The uranium itself was
regarded as an essentially worthless by-product, although a minute
amount was used in enamel glazes; the uranyl ion provides a luminescent green–yellow color to the glaze because of its ability to fluoresce.
The unused uranium was retained in waste stock piles, which actually
formed the basis for the first production of metallic uranium, both for us
and for the Americans.

The decisive step in purification in the process we had developed
at the Auer Company involved "fractional crystallization" of the nitrate.
We had gained a great deal of experience with this procedure since it
had been introduced by Auer von Welsbach (and subsequently used
by the Curies for the winning of radium) to separate the rare earths
and to concentrate both mesothorium and radium. P. Hoernes, an old
"keeper of the Holy Grail" of the Auer tradition and an early director of
our rare earth factory, whom I had "hidden" in our organization
because of his ethnic background[2], had developed much of the basic
work on the problems of uranium chemistry. Later on, as I shall relate
presently, we replaced fractional crystallization by another procedure.

We had no experience at Auer in producing metallic forms of the
compound. I employed colleagues at Degussa, our parent company,
for both reduction and melting. They had previously developed a
method for producing thorium metal that could easily be converted to
the production of uranium. Later on in Russia, we replaced their procedure by a much better one. The process of melting powdered uranium
into the cubic blocks desired in Germany proved to be very difficult for
us during the war because we were compelled to use a resistance-
heated vacuum furnace, not having a high-frequency unit. All such
endeavors were made especially difficult as a result of the shortage of
equipment and materials arising both from the other needs of war production and the bombings. As I recall, it took eight months to acquire a
transformer that contained 75 kilograms of copper. Copper was in
exceedingly short supply in Germany. As a result, we had produced
only a few tons of uranium fuel elements by the end of the war.[3] That
material was partly at Degussa in Frankfurt, partly at Auer, and partly
in the hands of the physicists involved in the development of nuclear
reactors.

---

[2]As is related in the oral interview and paraphrased in the introduction of this
book, Riehl shielded, in the Auer Laboratory, several partly Jewish colleagues
who would have fallen afoul of Nuremberg laws. He and his wife also shielded
the Jewish wife of a friend in their home toward the end of the war—a particularly dangerous period for retribution from the Gestapo as a result of the
attempted assassination of Hitler in July 1944.

[3]See the introductory material for further details.

One is frequently asked why the German uranium project never made more progress and why the Nazi regime did not support the program more energetically. The opinion is frequently given that the German scientists consciously or unconsciously did not wish to help Hitler's government obtain a weapon as deadly as the atomic bomb. This interpretation of the situation is not entirely wrong, but it is by no means the whole story. An investigator who had scientific curiosity or was interested in technical innovation could scarcely fail to be fascinated with such a development. The German program would undoubtedly have gone somewhat farther if there had been intense pressure and strong support from the government.

I believe that the relative laxity in the pursuit of the program rests primarily on the primitive level of intellectual understanding[4] of Hitler and his advisers. They had, perhaps, a good comprehension of rockets that roared with a great deal of noise and whose operations were apparent to them, but they had no real understanding of the unfamiliar, abstract concepts associated with the release of energy through nuclear fission. The lack of any requirement on the part of the government caused many of us not to expect any significant results from the uranium project before the collapse of Hitler's government. As a result, we never felt any compulsion to bring the matter to the attention of our consciences.

The atmosphere was completely different when we entered the Soviet Union. There we felt a strong undertow, stemming from the government, that reflected a brutally driven concern about the uranium program. Indeed, this was evident even before the explosion of the bomb at Hiroshima. All necessary personnel and materiel were sought throughout the country and made available for the program, often probably at the cost of other great needs of the nation.[5] Many scientific

---

[4]One of the provincial National Socialist leaders gained a great deal of international attention by making the remarkable statement, "Whenever I hear the word 'culture' I want to reach for my pistol!"

[5]The intensity of this drive, which continued well into the Gorbachev era, was appreciated by a substantial segment of the leadership in the United States but by no means all. Evidence obtained since 1990 has demonstrated, for example, that the Soviet military was working energetically on space-borne antimissile laser weapons, at least one of which was actually placed in orbit to test its possible effectiveness against missiles while in boost phase—the ideal period in which to destroy them. See the report by Robert Malloy, director of the Directed Energy Systems Division of Martin Marietta in Littleton, Colorado, which appears in *Space Trans.* (**1994**, 4(2), 1), the newsletter of the Space Transportation Association. While the U.S. Space Defense Initiative was being delayed by various forms of domestic opposition, including severe budget cuts, the Soviet

institutes associated with the Academy of Sciences or the various ministries were harnessed into service. Hasty instructions for the procurement of auxiliary equipment and material were given to many industrial organizations, often with threats of Draconian punishment if there were any miscarriages. Even the industries in Soviet-occupied Germany were put under pressure.

In the next chapter I will describe some particularly dramatic and even colorfully dangerous events that arose as a result of the pressures resulting from this atmosphere. The tension was made additionally high by our concern that the equipment and chemicals available to us were catastrophically limited. All that we had were the materials that we had stripped from our company and other places and brought to the Soviet Union. Even then, much was missing as a result of having been lost or damaged in transport. Missing, for example, was a large vacuum oven. I went to Zavenyagin, the Atomic Minister mentioned earlier, and wailed. He determined from a telephone conversation that it had inadvertently been shipped to Krasnoyarsk in mid-Siberia by mistake. A cargo plane was sent, and we retrieved it two days later. On one occasion Zavenyagin visited us in the tiny munitions laboratory where we were first located. He asked the staff of Russian workmen, who encircled him respectfully, from where the various pieces of equipment had come. The response was uniform. Each had been liberated as war tribute from Germany. Just as this exercise was finished, a rat suddenly ran by. He said harshly, "That clearly is ours."

To forestall misunderstanding, I should emphasize that in just a few years, the Soviet Union was producing electronic and other equipment of the highest quality.

We also had difficulties arising from what to us was the strange working style of the Russians. No Soviet Russian felt comfortable in carrying out a task without explicit orders from above. None were prepared to extend their area of responsibility even a small distance beyond their committed assignment. We were provided a good example of this during the first few days of our occupation of the new site. We called upon an electrician to connect our electric vacuum furnace that was to be used for melting. The process required the removal of the top of the oven, which was fastened with nuts on fixed bolts. Although a

---

program was moving ahead with customary intensity and boldness. The great significance the Soviet leaders placed upon their build-up of military strength is described in splendid, colorful detail in the four-part television series, *Messengers from Moscow*. Most of the narrative is provided by high-ranking governmental officials and military officers of the Soviet period. Their cynical attitude toward sincere Western peace and disarmament activities on the part of both private groups and government officials previous to the Gorbachev era is both dismaying and revealing.

wrench was readily at hand, the electrician refused to use it because that was the responsibility of a metal worker. Matters remained at a standstill for half an hour while we waited for a mechanic to arrive from the other end of the factory area.

The difference in work ethic between the Germans and Russians also led to frequent conflicts. On one occasion in the early stages of our work, a young engineer from Yakutsk came to me with a detailed list of questions concerning uranium technology. His approach seemed stupid to me, whereupon I became sufficiently annoyed that I left him and slammed the door. That evening I got together in our primitive housing accommodations with two German colleagues, Wirths and Thieme, and exchanged experiences. I described my encounter with the young engineer. Thieme said casually that it was not necessary to have gotten so worked up. It suddenly became clear to me that he was right. I stood up and went into the neighboring room where Golovanov, later our chief engineer, and another engineer were sitting. I asked them if they would kindly call in the young man from Yakutsk because I would like to express regret for my conduct. It was indeed very difficult to become acclimated to the Soviet structure and rules of behavior. Both Russians, who obviously knew of the incident, were so moved that tears flooded the eyes of the kind-hearted Stepanov. Thus, the incident ended with a marked improvement instead of a deterioration in our relationships. Moreover, I had made a friend of the young man from Yakutsk. Causes for outbursts occurred frequently later on. In most cases they were useful. They either helped us straighten out some difficulty quickly or helped melt the ice of relationships for the duration, after reconciliation.

There were no other plants in the Soviet Union involved in uranium production for nuclear reactors in 1945. We were the first to take on the problem. As a result, we began with the procedures we had been using in Germany. The wet chemistry processes and the metallurgical, or reduction, stage required the preparation of appropriate working space and the installation of large equipment. In contrast, the final step in fabrication, the melting of the metallic powder and casting of ingots, could be started at once because we had in our possession our German melting furnace and a small quantity of metallic powder that we had made earlier.

The preparation of the space and the mounting of the oven took place rapidly. One evening the whole German team assembled in the room where the melting furnace was located and spent the entire night assembling the equipment. We chose the night time in order not to be disturbed by visitors with their infinite number of questions. We left the Russian technicians out of the activity for the same reason. The members of the German group who were most intimately familiar with the technical details served as the guiding superstructure, while all others, including me, played the role of unqualified aides. The night time melting went somewhat less well than it would have under the best of circumstances. The activity had in some respects the characteristics of a symbolic, sacred ceremony, which indicated to the Russians that we were neither saboteurs nor dilatory. The director of the factory, the sympathetic general I mentioned earlier, called me frequently, and I provided "field reports" from the station where action was underway. The next day he told me that he had spent the entire night in communication with the offices of the Atomic Ministry (then the NKVD or the MVD), which transmitted the information to Beria who, in turn, transmitted it to Stalin. It bothers me a great deal even to this day to realize that we had disturbed the sleep of individuals at such a high level. Jokes aside, this reveals the authorities' intensity of interest and the degree of impatience in the program. They praised the fact that even the group leader would roll up his sleeves. Such action was regarded as uncommon in Soviet relationships. The leaders, particularly the politically schooled ones, limited their activities to giving advice and complaining.

The assembly was, in the main, completed by the end of 1945. Production proceeded slowly. On the German side, Wirths was involved with the activities in wet chemistry, whereas Ortmann, who actually was not a metallurgist but a long-term colleague of mine once involved in luminescence, followed the work in metallurgy. (In Chapter 13 I will attempt to explain why essentially all the German scientists were able to switch their field of interest rather easily to one that was relatively foreign to them). Our production, however, remained far below what the government desired. The atmosphere became continually worse and more tense, leading to an unpleasant scene that I will relate in the next chapter.

We gained a small transient amount of prestige in the beginning of 1946 when, in a few days, we produced a ton of reactor-pure uranium oxide in the form of spherical pressings for the benefit of the Soviet physicists who needed it for a major investigation. This was another instance in which I set the entire German team at work, both day and night.

Our failure to maintain targets for production was not a result of the quantity and size of our equipment but in the technology of fractional crystallization that we were using. It was very good for purification but required time and formed a bottleneck in the chain of steps leading to production. We learned of the possibility of using a better method in the following way.

Soon after the explosion of the bomb at Hiroshima, the American government released a book written by Henry D. Smyth that described the development of the atomic bomb. The book was immediately translated into Russian and distributed to all projects. I also received a copy. I worked my way through the book in a single night. It mentioned briefly that the Americans had purified uranium by an ether process. This involves covering or agitating an aqueous uranyl nitrate solution with ether. The uranyl nitrate dissolves into the ether, leaving essentially all impurities behind in the aqueous phase. The technique was well-known to us but we had used it only in test work for enriching the amount of rare earths in the aqueous solution for purposes of analytical determination. Because it was necessary to discuss ways of replacing fractional crystallization by some other method that would give a higher through-put, I described to my co-workers the method the Americans had used on a large scale, in spite of the hazards of fire associated with the use of ether. We said to ourselves we can do anything the Americans can. Wirths and Thieme worked out the process to scale very rapidly. The operation had to be carried out in a leak-proof reinforced system in order to avoid the danger of an explosion derived from ether vapor (today one uses tributyl phosphate instead of ether). We obtained the necessary ceramic containers, tubes, and flanges from the Hermsdorf ceramic factory in Thuringia. As a consequence, the "ether enterprise" was ready to run by June 1946.

At this point, however, the Russians got cold feet because of the dangers of an explosion. We failed to get the necessary permission from the superiors in charge to start the operation, in spite of our numerous requests. Finally, we became annoyed and I went to Moscow to see Zavenyagin for his help in setting the process in motion. He was away, so I saw B. L. Vannikov, with whom I also had very good relations. Vannikov had been minister of munitions during the war. He was, at least at that time, situated a bit higher than Zavenyagin. As a lieutenant general, he also had a higher rank than the latter. I said that, in as much as we had been so severely criticized for insufficient production, we did not understand why we now had to hold back when we had a really good process. After further persuasive discussion on my part, Vannikov gave us permission to proceed with the ether process. The production rate jumped up immediately. The yield of the ether-based equipment reached nearly a ton per day. It operated for many years without incident, until the process was finally replaced by an entirely safe one.

Years later I noted that my appeal to Vannikov was resented by Zavenyagin. Once, during my absence, he complained to professional colleagues at a technical institute during a discussion of indecisiveness: "Take as an example, Nikolai Vassilyevitch (my Russianized name). When we were worried about setting the ether factory in operation, he took an occasion when I was away to get Vannikov to give the

Henry D. Smyth, author of what is called the Smyth Report, Atomic Energy for Military Purposes, which proved to be useful to the Soviet teams. (Courtesy of the Emilio Segrè Visual Archives of the American Institute of Physics.)

orders to do so." I did not contradict this somewhat incorrect interpretation of the situation, for who does not like to be pictured as a daring adventurer.

During the meeting with Vannikov, I learned something that may be of interest to physicists in regard to the fate of the well-known Soviet physicist, Peter Kapitsa. An individual suddenly came into Vannikov's office to say that Kapitsa and all his colleagues had been relieved of their positions. It was offered as a sensationally sensitive matter and was clearly a surprise even to Vannikov. Years later, I learned that in falling out of favor Kapitsa was without a post for many years and carried on limited physical experiments with his son in a shed of his dacha under cramped circumstances. He was "rehabilitated" only after the deaths of Stalin and Beria and then was treated reasonably well again.[6] I could not determine the precise reasons for his demotion by Stalin. It appeared that he was unwilling to work on the atomic energy program. I had met him and the distinguished physicist A. F. Joffe[7] in 1945 when they had attended a lecture on uranium technology that I gave in Moscow before a prominent group of physicists. Both came to meet me after the lecture but showed no interest whatsoever in uranium. Joffe was not an expert in nuclear physics and was primarily interested in my work in the field of luminescence, whereas Kapitsa asked about the fate of several German physicists, especially about Walter Gerlach. I gained the impression that he was openly displaying his lack of interest in the uranium program.

After 1955, someone showed me a book with the title, *The Red Atomic Czar Kapitsa*. The author, a Russian who clearly had defected to Germany during the war, mentioned several physicists who were a part of the atomic project later on, for example, Kurchatov, but everything else in the book was an extrapolation and a fantasy. In any event, after 1945 Kapitsa had nothing to do with the project. The leading physicist involved with it was I. V. Kurchatov, with whom I had a great deal to do because he was the "customer" for our uranium. As is well-known, Kapitsa received the Nobel Prize and was even allowed to leave the country occasionally. Mutual friends tell me that in spite of his

---

[6]At a meeting at the Ettore Majorana Centre at Erice in Sicily in the summer of 1983, Paul Dirac gave a brilliant and moving account of his long relationship with Peter Kapitsa. Kapitsa had been allowed to attend a meeting at the center at that time. Dirac includes a dramatic incident that occurred at Kapitsa's dacha on the day Beria was assassinated. Special agents were assigned to protect him from any possibility of revenge by Beria's followers. Dirac's lecture is available at the Erice Center.

[7]In his biography von Ardenne mentions visits Joffe made to his laboratory in the Soviet Union.

General Boris L. Vannikov, left, General Zavenyagin's superior in a relaxed meeting with I. V. Kurchatov in 1959, after Vannikov's retirement. Vannikov was an expert in industrial production and a genial friend to those who knew him. He reported directly to Beria before the latter's death. (See the organization chart on page 78.) (Courtesy of R. Kuznetsova of the Archives of the I. V. Kurchatov Memorial Institute.)

age he enjoyed mountain climbing. His death was announced recently (1984), as this book was going to press.

Having increased the production of uranium as a result of the ether process, our bottleneck had become the "hot" process, that is, the production of the pure metal. This portion of the program also caused us a great deal of anguish. The use of calcium as a reducing agent did not generate a bolus of pure uranium metal. Instead it produced a mixture of uranium powder and calcium oxide. The calcium oxide could be removed by acid, but one then obtained uranium powder with inclusions of oxide.

A postwar photograph of Peter Kapitsa presented to Frederick Seitz during a 1969 visit to The Rockefeller University.

Abraham F. Joffe (left), one of the leaders of Soviet physics before, during, and after World War II. He avoided working on the nuclear bomb
program but stayed in contact with Manfred von Ardenne in Sukhumi
working on other basic scientific work. The other two scientists are
P. Kapitsa and A. N. Krylov. The photo was taken in Paris in 1928. (Courtesy of the V. Ia. Frenkel Leningrad Physico-Technical Institute Collection
of the Emilio Segrè Visual Archives of the American Institute of Physics.)

One day I received a visit from the Platinum Colonel mentioned in
Chapter 2. He asked me why I insisted on the oxide process. I said that
we had not insisted on it but had no other method with which we were
comfortable. He persisted further and expressed the opinion that one
could use uranium tetrafluoride instead of the oxide. The calcium fluoride that would be generated could be melted along with the uranium,
in contrast to calcium oxide, and permit the formation of liquid uranium
that would produce an independent uranium bolus at the bottom of the

crucible. I responded somewhat sullenly to the effect that I had also suggested such an approach, but that we had felt committed to the oxide process. Gradually however, I realized that he knew a great deal more than I did and began to listen to him. As a result of a form of groping through the valuable comments being made by the Platinum Colonel, I came to realize that he was trying to put me on the right path without giving me the actual source of his information. I am now certain that this information had been gained from America through espionage.[8] Later on I obtained more direct indications of the results of espionage. In my presence, a high official in the Atomic Ministry asked a deputy about the quality of our uranium metal. The answer was, "It is even better than the Americans'." The Soviets had succeeded in procuring and analyzing a specimen of the American metal.[9]

The fluoride process, which we pushed ahead at once, proved to be far superior to the oxide route. The head engineer of the factory, Golovanov, and Wirths carried out the technical development of the process very effectively. The final step in the production of the uranium fuel elements also fell into line. The remelting and casting of the uranium was handled successfully with the use of the induction-type vacuum furnace we had procured in the meantime. Our German group did not play a role in the cladding of the fuel elements with aluminum.

---

[8]Riehl's surprise at the extent to which the Soviet espionage system had penetrated the United States produced no such surprise to those of us who had been linked to the Manhattan District, but not because of outright carelessness on the part of security staff, which was diligent. The sites of the Manhattan District were large and spread out in the country. Moreover, Soviet representatives had extensive access to all parts of the nation. The only surprise might have centered around the question of "who?" It is significant that none of the major scientists in the operation of the Manhattan District who cooperated with the Soviet spies were U.S. citizens, although the individual whom the Soviets termed "Perseus" and whom they will not identify is still unknown. The claim by Pavlov and Anatoli Sudoplatov that J. R. Oppenheimer, E. Fermi, and L. Szilard cooperated in espionage, as reported in the book *Special Tasks* by Jerrold L. and Leona P. Schechter (Little, Brown and Company: Boston, MA, 1994), is sheer nonsense.

There is a story, perhaps apocryphal, that a small group of American scientists, gathered together for a relaxing evening during their period of most intensive work at Los Alamos, New Mexico, raised as a sporting game the question: "Who among the scientists here *might* be a spy?" The agreement reached in a light-hearted way was: "Klaus Fuchs!"

[9]In the televised documentary (see footnote 2 of the introduction), Wirths mentions that when his Soviet leaders praised him for the higher purity of the product of his group, he responded, "We are probably over-doing it. The Americans have probably picked the optimum purity to gain the advantage of quantity production." One of the Soviet leaders responded, "You damned Germans!"

Igor V. Kurchatov during his student days in Baku in the 1920s. He was obviously hard-driven even then. (Courtesy of the Emilio Segrè Visual Archives of the American Institute of Physics.)

That was accomplished by the aluminum experts in the aircraft industry without encountering significant obstacles.

Nothing further stood in the way of the rapid development of a system for manufacturing uranium metal for the Soviet reactors. The tension was reduced. By the time we left Elektrostal in 1950, the factory was producing as much as one ton per day of fuel elements ready to be placed in the reactors. Needless to say, by this time ours was not the only factory.

We learned about the explosion of the first Soviet nuclear bomb in 1949 over the British radio, which we listened to every night. I went to Golovanov the next morning and told him the news, which was not generally made available to the Soviet public.[10] He departed at once

---

[10]It is an interesting fact that Soviet officials such as Golovanov did not systematically listen to British radio as Riehl did. It is not clear whether this was a result of a language barrier or fear of criticism from the Party bureaucrats.

and went to Moscow to obtain confirmation. The news was announced by TASS several days later but in a confused, watered-down version in which it was said that the Soviet Union had had the secrets of atomic energy at its disposal for several years. Nothing was mentioned about bombs. Emphasis was placed upon the peaceful uses of nuclear energy for such matters as diverting rivers. The announcement breathed a message of technical sovereignty and an essentially relaxed, moving love of peace.

Not long after the successful explosion of the bomb, honorary orders and special prizes poured forth for us. Even the German group received a shower. My German colleagues, Wirths and Thieme, received the Stalin Prize and the Order of the Red Flag. I received the largest award. In addition to the Stalin Prize, I received the title "Hero of Socialist Work," with the Golden Star, which, when worn on my breast, stirred looks of wonder on the faces of many top-ranking politicians. (In the case of military personnel, a similar golden star carried the designation "Hero of the Soviet Union.") Along with this I received the Lenin Prize as well as the gift of a dacha in a beautifully wooded region west of Moscow where government officials had their special residences.

Igor V. Kurchatov in his office in the period following the success of the initial phases of the Soviet nuclear bomb program. (Courtesy of the Emilio Segrè Visual Archives of the American Institute of Physics.)

President Nikolai M. Shvernik of the Soviet Union and chairman of the Presidium of the Supreme Soviet presenting an award, presumably to an outstanding Soviet worker. Riehl received his medals and other very high honors from Shvernik and was, as a result, invited to special functions at the Kremlin. (Courtesy of the Russian Pictorial Collection of the Hoover Institution Archives.)

My Stalin Prize carried a large amount of money with it. I said later to Zavenyagin that I had never previously been a capitalist and was astonished to find myself one in a socialist country. The heaping on of honor and wealth was indeed a heavy burden for me. My wife was horrified and expressed the opinion that we would never get out of the Soviet Union. I, however, retained the will and hope that we would indeed get away.

Chapter 4

Comments on Typical Incidents

In this chapter, I would like to relate a few typical events and situations that illustrate the tension-loaded atmosphere in which we worked in the course of the uranium project and describe the pressures placed upon us by the government,particularly during the early period.

We required a thick-walled rubber vacuum hose of the type used in industry for our melting equipment. I went to Atomic Minister Zavenyagin and asked for his help. He promised to make an official government request to the two factories that manufactured rubber galoshes ("Provodnik" and "Treugolnik") to manufacture such rubber tubing as fast as possible. After a time, we received several hundred running meters of hose. It was fine with respect to exterior appearance, being uniformly circular in shape. The inner profile however was truly grotesque. It was completely irregular, having the aspects of venturesome avant garde art produced by an imaginative artist. It was, however, completely useless for technical purposes.

During this period, I was having a small war with Zavenyagin who was attacking us for our low rate of production, and I was defending our group on the basis of the difficulties we were having in getting materials and equipment. The miscarriage of this request for rubber tubing provided me with an excellent defensive weapon in our battle with the administration. I cut out a segment of the hose with a particularly amusing profile and went to Zavenyagin with it. Several individuals were visiting him in his office as I entered. I said, "You have continuously assured us the support of Soviet industry. May I give you an example of the quality of production of your industry." Then I laid the sample of hose on the table with ingeniously planned deliberation. Zavenyagin erupted in anger. He said curtly, "Can I keep this sample?"

I handed it to him with a smug smile and paid my respects. Later on, I heard that the unfortunate director of the galosh factory had received a five-year jail sentence. Thereafter, I displayed great hesitation when there was any justified criticism of anything or anyone.

Another incident also had to do with Zavenyagin. In order not to appear to put him in a bad light, I would like to make an excursion beforehand to characterize the man. In 1945 Abram Pavlovitch Zavenyagin was made Atomic Minister, along with Kurchatov, the physicist who was placed in charge of the entire atomic energy project. Previously, Zavenyagin had a major role in the development of an enormous nickel combine on the Tamyrian Peninsula under the suzerainty of the NKVD. His previous experience had been in the field of metallurgy. Although his first name was Jewish, his ethnic background was not. Moreover, he was of the physical type found commonly in the Volga region of central Asia among the Volga Tatars, who had once formed the Islamic "Bolgar Empire." These often handsome, brunet individuals exhibited Mongolian facial characteristics only rarely, resembling more closely the modern Bulgarians. They constitute a valuable, industrious element in the mixture of populations in the Soviet Union. Tatar descent often carries with it a state of nobility and an impressive personality, as well as competence in the field of the exact sciences. The designation, Tatar, or Turko-Mongolian, does not connote a sharply defined ethnographic grouping. The Russians designate most of their enemies east or southeast as Tatars. This includes the Crimean Tatars who sustained a highly evolved Islamic culture in Crimea for centuries. Many outstanding personalities emerged from this group after Russianization. It should be added that I. V. Kurchatov, who was born in Crimea, had a quite definite Tatar appearance.

Zavenyagin was a particularly capable and wise person. His speech was extraordinarily brief and succinct. His rough surface tended to mask the fact that he was basically a courteous and refined individual. He showed a high level of good will toward me. However, he compensated for his high regard for German science and technology through needle pricks directed toward me in which he denigrated German technology. I promptly reacted toward each provocation in a sour manner. It is quite natural for one to complain about domestic matters when at home in one's own country but to defend them heroically when abroad. I must state here that I felt pangs of conscience toward Zavenyagin when I left the Soviet Union in 1955. I doubt whether he fully appreciated the enormous value of personal freedom and the basic reason why I wished to leave. In later years, I noted a considerable similarity between Zavenyagin and Gorbachev with respect to features as well as behavior.

A strange characteristic of Zavenyagin, which I am about to discuss, was his custom of using the most vulgar, foul language. The Russian language is particularly rich in this respect. I have been told that this means of expression was developed in the years of the "Tatar Yoke."[1] Moreover, Zavenyagin and his colleagues, particularly those in the army and industry, cultivated this habit during the stresses of the war years. The Russian curses and expressions of complaint are so vulgar that the fecal language of the Westerner seems like an elevated style of conversation in comparison.

We found ourselves in the deepest rut of our troubles in January 1946. Nothing went right. Nothing showed signs of improving. The mood was dreadful. I was out in the open country around the factory one day when the director of the factory, the sympathetic general, flushed me out and asked me in an excited voice to come to the head office. He said that Zavenyagin was coming to visit and, making a grim face, he said that we should expect a "*mordoboy*." *Mordoboy* is a non-literary expression that one uses to designate very nasty abusive behavior. It originates in the Russian word for slapping one's ears or perhaps more in the spirit of "socked in the mouth." (This type of familiar language is encountered only in Soviet industry, never in academic circles.) We went into the mess hall where 20 or so of the individuals involved in the development of the factory were seated. I sat behind Zavenyagin and took several drinks of schnapps on an empty stomach in order to be prepared for the battle. After the meal, the entire group went into the head office and the "*mordoboy*" began. (Someone told me later that I quickly sobered up as matters became serious.)

The quality of the disciplinary criticisms we received was beyond Western standards as Zavenyagin directed his words, one by one, to the entire group: the director of the factory, the head engineer, the NKVD general in charge of the construction workers, the leader of the group directing the program from Moscow, the division leader in charge of procurement, and so forth. It was clear that I was to be included in all of this. The tone of the criticism was unbelievable. Zavenyagin addressed the individuals in the most vulgar manner of which the Russian language is capable, irrespective of rank or age. The mildest of these curses would have been judged a major insult in any Western democratic country. The level of criticism encompassed both the flushed and the pale, and no attempt at justification was made. I would not have put up with this personally under any circumstances. I would have quietly stood up, gone slowly to the door, and left the room slamming the door behind me. When it came my turn, however, he said, "As for you Dr. Riehl, I can say that the level of opinion

---

[1]The period of the Tatar Yoke extended over three centuries from the thirteenth century, when the Mongols invaded Russia, to the time of Ivan the Terrible in the middle of the sixteenth century, when the Muscovites gained the upper hand. The Russian victory in the Battle of Kazan (1552) was the turning point.

held of you in Moscow has definitely fallen." I responded to this mild assertion with the equally mild statement that our support had been deficient. After an undramatic verbal skirmish, he let me alone. I do not know whether he was treating me lightly as a foreigner or because he recognized my resolute reaction to each stab or insult to the others.

The individual criticized most severely was an old gentleman named Feodorovitch, a one-time Czarist guard officer who was director of the planning group. He often provided me with good advice, particularly with the advice to remain behind anyone, or at least out of the direct view of anyone, offering criticism. One can, in this way, often avoid such criticism personally. I could not follow his advice on this occasion because I was seated directly opposite Zavenyagin. Feodorovitch sat on a sofa along the wall, immediately behind him. Unfortunately, that did not help him on this occasion and he took his share of criticism. As the blasting came near its end and the tension began to abate, Feodorovitch smiled at me from the sofa. Because Zavenyagin was a sensitive observer, he discerned from my response that Feodorovitch had smiled at me. He turned abruptly and said to him, "You have no need to laugh!" With that poor Feodorovitch received a second dressing down.

A similar incident, in which the criticism occurred in a more harmless form but in which the issue was very serious, occurred considerably later when our production was running well. Suddenly, much too high a concentration of boron, the number one enemy for a nuclear reactor, began to appear in our uranium metal. On this occasion, Vannikov came to us to apply pressure. That was the last time I was to see him. I suspect that, as a former minister of munitions, he subsequently took on a new assignment that was regarded not to be of any concern to us.[2] None of us knew where the boron originated. The tone with which Vannikov put the pressure on us was not uncourteous but threatening. For example, he asked the engineer-in-chief whether he had ever had meals in the Lubyanka Prison, the famous correctional facility of the NKVD. When the paled engineer answered affirmatively, Vannikov said, "Would you like to go there again?" A possible solution to the excess boron finally occurred to me and eased the mood. I remembered that the uranium oxide that we had produced in the Auer Company had been stored in a shed that had previously contained boric acid, which was required in the production of luminescent materials. Back in Germany, the zealously active service officers of the NKVD had scraped together all the uranium oxide on the floor of the shed along with some dirt. It was possible that the presumably contaminated uranium oxide had been used as a starting material for our production.

---

[2]General Vannikov was probably retired, but with high honors, following Beria's execution as a result of being a close associate of the latter.

Although the ether process is excellent, it was not capable of removing unusually large quantities of boron. The boron problem eventually vanished, and the incident had no bad consequences.

In concluding this chapter, I would like to add a few words about the tension-laden atmosphere of the time and about the working style of the individuals I knew in the program, especially those who were involved in particularly strenuous programs. They were all under very great stress, both during and after the war. Almost all of them developed heart trouble. Zavenyagin and Kurchatov died of heart attacks. Vannikov complained of heart problems but was still alive in 1957. The working practices of these men were very unhealthy. Zavenyagin had many of his most important conferences, which I unfortunately had to attend, at ten o'clock in the evening. On those occasions, I usually returned to Elektrostal at four o'clock in the morning. I could never sleep in the automobile going back because of the exceedingly dangerous state of the road. For a period of time during the war, the highway was the only link with the east. It was, incidentally, the first part of the famous, indeed notorious, Vladimir stretch over which prisoners destined for Siberia were driven in the previous century. In the 1950s the night time meetings were abolished. Also, the use of foul language during work was forbidden. This change must have gone very badly with Zavenyagin.

Chapter 5
Two Encounters with Beria

I had two encounters with Beria, the infamous organizer of the NKVD work prisons. The first relatively simple meeting occurred soon after we were taken to the Soviet Union. Beria had invited Hertz, Volmer, von Ardenne, and me to visit him in order to become acquainted with us. Each was separately invited into his office where perhaps 20 other individuals, mainly scientists and a minister, were seated.

Beria greeted us very cordially. His approach was exceedingly pleasant. It is well-known that people regarded his manner in personal matters to be very nice. I am told that even Himmler could be a charming companion. At the very beginning of our chat he said that our people should forget that we have been in a terrible war with one another. He had the opinion that the Germans were very proper and would display respect for orders when given. He stated, by way of example, that he had learned that if, at the end of a battle, no explicit orders were given to stop shooting the German soldiers would continue firing.

He proffered the following joke about the Germans. The Germans stormed a railroad station. The conflict came suddenly to a halt. The commanding officer sent his adjutant to see what had gone wrong. He returned and said, "There is no need to worry. The forces are acquiring their train tickets."

The remaining conversation contained little of note. Outstanding was the curiosity with which Beria was observed by all present. Particularly noteworthy to me was a man with a dark beard and glowing black eyes who watched me with an unbroken friendly look. Later I learned that he was Kurchatov (see photograph on page 103).

The second, more interesting, encounter occurred about three years later at a time when our technical difficulties were essentially behind us. One day I had a very bad cold and decided to remain at

Lavrenty Beria, the notorious and greatly feared head of the intelligence agency, who was assassinated by his colleagues soon after Stalin's death in 1953. (Courtesy of Photo Novosti RIA, the Embassy of the Russian Federation in Washington, DC.)

home. The telephone rang, and the director of the factory was on the line. He said that he knew that I was sick, but he begged me to come to the works. I replied that this was the first time in three years that I had stayed home because of illness and that I would appreciate being left to rest. He did not let go. It would be an exceedingly important visit, and it would be dreadfully improper if I were not present. After a long dispute, I finally realized that my refusal would put the management in a very unpleasant position, so I decided to go to the factory.

I waited in my office and looked out the window from which I could see the entrance to our laboratories. After a period, a virtually endless caravan of large black limousines arrived. There were at least 15 cars. Beria climbed out of one of them. I realized then that this really was an important visit and went out into the corridor to meet him. He asked, "How are you?" I replied, "Terrible, I have the grippe." He replied that he knew I had the grippe and would send me a special remedy that he had. (I have not yet received it to this day.) We went into my service room, which was filled with people, some sitting and some standing. Present were a number of ministers, the director of the factory, the party leader of the area, and many other individuals unknown to me. Several huge, strong young men were posted around the walls of my service room and in the entrance chamber.

Before going further, let me say that I was in a very dark mood at the time. I not only had a bad cold, but because of it had to forego my customary cigars, which otherwise would have calmed and "moderated" me. Without the assistance of cigars, I tended to get into a state in which I was inclined to be alert, tense, and too aggressive. (I tried to give up cigars many times during my lifetime, but I always became obdurate and pugnacious as a result.) As a result of this, chatting with Beria produced one of the most amusing incidents in my life.

The situation was peculiar from the start and not devoid of comic aspects. One sensed the way in which everyone around Beria trembled. Even the much feared Zavenyagin became "small and mean." Only I, the "focus" of the event, had no obvious reason to be afraid. Beria had not come to visit us for any special reason having to do with me. He was serving the needs of his colleagues. Moreover, the good progress we were making in our work gave assurance against any danger. Thus, it came about that on this occasion I did not feel threatened and, in view of the submissive, worried mood of the others who were present, I found the situation somewhat entertaining.

Beria opened the discussion by asking what we were up to now and how things were going. I reported briefly on our ongoing work, which no longer dealt with uranium metal but with uranium 235 and plutonium. However, the presentation awakened no interest in Beria. After a while he asked whether we had some complaints to make. As a consequence, I brought up a really harmless hardship, which I cloaked in the form of a well-known saying from Russian history. Russian history relates that the Russians went to the "Varangians" and said, "Our land is large and rich but there is no order in it. Please come and govern us." I said, "Our land is large and rich, but we have no pure chemicals." Beria smiled as a result of my jest-like formulation, but no one else departed from their earnest and respectful stance. The minister of the chemical industry, M. G. Pervukhin (later ambassador in East Berlin and still later a member of the Central Committee, ZK), who was sitting next to Beria, smiled least of all. Beria looked at him questioningly, and Pervukhin replied that the problem was well-known and that he was establishing a special management section in his office devoted to the area. With that, the theme was dropped.

General Michail G. Pervukhin, minister of the chemical industry at the time of Riehl's second meeting with Beria, after the successful production of uranium. (Courtesy Photo Novosti RIA, the Embassy of the Russian Federation in Washington, DC.)

Beria said that it was unlikely that I had only a single complaint. I fished around my mind and found another, saying that the lack of high-temperature crucibles in the Soviet Union was a hindrance to our work. I obtained even less response from Beria as a result of this complaint than I had regarding the previous one about chemicals. He pressed me further, and it became clear that he was searching for a deeper, more painfully bothering complaint. He became more explicit by saying that I had only raised issues related to our professional work, but there must be more personal issues about which I and the members of the German group had complaints. I said, speaking in a cold, sharp tone, "We have enough to eat, we do not freeze. We have no complaints." To appreciate the nature of my response, the reader must understand that each privilege or benefit the Germans received would entrap us further in the Soviet net. That had now become evident to me, and from the start I struggled to get out of the net. As a result, it was best not to ask for anything other than matters absolutely essential for the sustenance of life, particularly those related to health. Beria said, "That is unlikely. Everyone has something to wail about." He pressed me further, and I finally said, "If you insist I complain about something, I will do so. I complain about you!" The effect of this comment, which I am giving essentially word for word, was remarkable. Those surrounding Beria were petrified, whereas he was apparently amused and with feigned fright asked, "About me?" I said then that he had personally ordered rigid maintenance of secrecy and the security system, that our freedom was monstrously curtailed, and that we suffered under such conditions. Beria began to deliberate among his colleagues as to whether an exception could be made for our group. However, I waved the matter off and opined that I had only brought up the topic because he had pressed me. I asked nothing of him.

When I later described this to my German colleagues, none of them reproached me, although I had a feeling for the way in which they gnashed inwardly. There were numerous grounds for avoiding additional benefits: First, we knew from experience that our "freedom" would be curtailed even further; second, there was the high degree of suspicion that would fall upon our group if there were any leaks of secrecy; and last, but not least, one did not want the acceptance of any privileges to cause difficulties with our return to our homeland.

Additional discussions with Beria subsided in a way I cannot recall. The group broke up to visit the factory. Zavenyagin wanted me to join them, but Beria waved him off saying, "The man is sick. He should go to bed." Zavenyagin stood somewhat behind the group and expressed his personal thanks to me in the most effusive way I could imagine. I had no idea why he expressed such gratitude.

I had probably failed to comprehend the deeper purpose of the entire event and the notable questioning by Beria. Much later, someone explained to me the reason for the visit and the questioning. The Soviet scientists, particularly those in the Institutes of the Academy of Sciences, were accusing Zavenyagin of preferring to obtain advice from the Germans instead of from them.[1] This reaction was not unreasonable, for there were excellent scientists in their organizations. This complaint made it possible for Zavenyagin to arrange to show Beria that he had a very productive German group under his control and that this justified his actions. Clearly my presentation and what went with it conformed to Zavenyagin's intentions and hence, his overwhelming thanks.

---

[1]Doubtless Beria found it convenient to work with German captive scientists because he could place them under restrictions that would have been difficult to apply to Soviet counterparts of equivalent standing.

Chapter 6

Beria Once Again

In 1953 we were in the last of the places we would live during our residence in the Soviet Union, namely Sukhumi on the Black Sea. I was sitting in my study when my German secretary rushed in and asked whether I had a picture of Beria hanging in the office. It became evident that Beria had been removed as one of the three or four leaders following Stalin's death. It was sensational.

During the next day or the day following that there were reasons for having special meetings in the entire Soviet Union. They were termed "People's Gatherings." The meeting was held in the open in our compound because the weather was very warm in the south. (The compound consisted of the area surrounded by barbed wire, which enclosed the institute where we worked, the residences of the Germans, and a part of the residences of Soviet colleagues.) The Germans were not invited to the meeting. I was, however, overcome with curiosity and took a position behind a fence where I could hear everything and also see a bit without being seen. The participation in the meeting was small: Less than 100 individuals were present. One felt something in the nature of embarrassment and lack of security in the overall situation. One had to consider the fact that Beria was a prominent figure, the most feared among the top group. A few of those present were openly pleased. They were in an elevated mood derived from a mixture of malicious joy and love of the sensational.

The principal talk was given by the deputy head of the compound. (The head had gone off to avoid the obligation.) Inhibited and without glancing at his manuscript, he droned on his accusations against Beria. The accusations were so unbelievably crude as to be beyond imagination. No mention was made of his concentration camps or of similar horrors. There was mention of the immoral acts he committed through misuse of the power of his office. Above all, however, it was claimed that he had betrayed his country. He had worked in close association with the German General Staff since 1919!

It is difficult to avoid making a peripheral comment at this point. Beria was 19 years old in 1919. He became a traitor very young! What is additionally astonishing is that this treachery was not discovered by Stalin during 34 years and that Hitler had not used it to drive straight through to Vladivostok. What negligence on the part of these noble dictators!

The last speaker asked the assembled people what the penalty for Beria's crime should be. The group shouted, "Death!" It was a safe guess that this penalty had already been meted out. In a People's Democracy, the People's will is frequently carried out before they have expressed it. Revolted by the proceedings and somewhat ashamed for having witnessed it, I stole away.

Beria was born in a large village not far from Sukhumi that we passed frequently on our way to the mountains. He was not a Georgian but originated in the Mingrel community that dwells in Georgia. A large gypsum statue of the great son of the village stood on the edge of the village. The statue vanished after his downfall, but the pedestal remained. The village had the hope that it would someday produce another great son to commemorate. May his statue stand longer, if and when he comes!

Chapter 7

The Good Russian Man
(Mensch, Dostiny Chelovek)

"The good Russian man" is a common expression. It does not mean that all Russians are good without exception. Rather it describes a characteristic that is often found among Russians. It is an expression similar to "the golden Jewish heart" or "the loyal Teutonic soul." The best formulation of this quality that I have encountered is that by the German Max Frisch who said, "When the Russian is not an ogre, he is more human than we are."

When a Russian gives you advice with the assurance that it is that of a good Russian, you can be certain that, as far as it goes, the advice is sincerely intended and well based. I knew that and took advantage of it. The first event concerning this characteristic occurred during the initial period of our residence in Elektrostal. I was tearing out my hair in connection with the work on the uranium project and hoped within my heart that I could escape to a less taxing activity. The Russian engineer Stepanov, mentioned in Chapter 3, observed this and came to me one day. He said with compassion, "Listen to the advice of a good Russian, do not become ruffled." I lowered my high level of vexation.

A similar case occurred to me some years later. A young German girl in our group wanted to take a trip to Moscow. This required that she be accompanied by one of the escorts from the security forces without which we could not leave Elektrostal. The event occurred on a Sunday, and I could stir up only one escort, a young lad. He, however, refused to go to Moscow. When I asked him for the reason, he answered without any shame, "Because I don't want to go." I threw him out in anger and went to the director of the factory the next morning and requested the immediate dismissal of the lad. I was promised that this would take place at once, but nothing happened after several days in spite of my repeated insistence. I finally went to an elderly sympathetic NKVD colonel to whom the escort reported directly. He evidently had known of the incident from the beginning, indicated full appreciation, and said, "Please understand. This young man writes reports concerning you, me, and the director of the factory. Take the advice of a good Russian and drop it." I gave up. Some two years later, after we had left Elektrostal, I learned that the young man was dismissed for stupidity and arrogance.

In the third case, the actions of the good Russian were presented not in the form of words but as a noteworthy deed.

Each German family was permitted to send one package a month of the best food to a relative or a friend in Germany. The package was taken to Berlin by means of the NKVD (or MVD), and it was sent on from there even to West Germany. Baroni, an Austrian in my group, could not send packages in the same way to his father in Vienna because the agreements with Austria prohibited any direct transportation through the services of the NKVD. I gave a good deal of thought to ways of getting around this problem but had no success on my own. As a result of this glitch, Baroni's packages were stored in the home of the officer in charge of the shipments. I recall the full name of the officer but, for the sake of camouflage, I will call him "Ivanov." One day my "spy" informed me that the mound of packages in Ivanov's home had vanished. A gloomy suspicion crossed my mind. Because I was nominally Ivanov's boss, I called him in and in the very formal manner of a chief of office, I said, "Where are the packages of Dr. Baroni?" Ivanov shifted from one leg to another and kept repeating the apparently meaningless sentence, "Yes, where are the packages of Dr. Baroni?" Finally he took hold of himself and confessed, "I will tell you, but if you do not keep it to yourself I can obtain a 10-year prison sentence for breaking official rules. As a Soviet officer, I have special methods for sending packages to Vienna, and I have used them. The old man (Baroni's father) should not starve." I abandoned my formal posture.

I never told this story even to my wife, keeping it secret until our return to Germany. The packages were sent to Vienna in the same way two more times with my assistance. I stood by the entrance of the post office as a lookout so that none of the Russians or Germans who knew us would see what was going on. Finally we found a legal way to do the job.

Chapter 8

In Sungul in the Urals
(1950–1952)

By 1950, our work at Elektrostal was completed. The production of uranium was proceeding smoothly and no longer needed the cooperation
of the German team. However, the time was not yet ripe for a return to
Germany. A preliminary attempt on my part to achieve this was unsuccessful. Thus, the question arose as to what to do with us next. Zavenyagin proposed that I take over the directorship of an institute in Sungul,
east of the Ural Mountains (see figure on page 84), that would be
devoted to the handling, treatment, and use of radioactive materials
produced in reactors, the "fission products." Related to it would be radiation biology, dosimetry, and radiochemistry, as well as problems in
technical physics. In other words, it would be an exceedingly many-
sided institute. I had gained experience and confidence in more or less
all of these areas during my years at the Auer Company. Hence,
Zavenyagin's proposal seemed to be justified and even enticing.

My decision to take the offer was made easier by the fact that three
colleagues with whom I had very close and friendly relationships were
already at the institute. They were the physicist and radiation biologist
K. G. Zimmer, mentioned in Chapter 2; the radiochemist H. J. Born, a
student of Otto Hahn (later a professor of radiochemistry at the Technical University of Munich); and the physician and radiation biologist A.
Katsch, who would later become a professor at Karlsruhe. Although
these individuals had been close to the Auer Company, they actually
had been on the staff of the Kaiser Wilhelm Institute in Berlin-Buch and,
indeed, in the section led by the geneticist N. V. Timofeyev-Ressovsky,
about whom more will be said presently. The Russians brought these
scientists to Elektrostal and initially placed them in my group. It was,
however, very difficult to do justice to them professionally within the

framework of a uranium factory. I tried frantically at the time to prove
that radiochemistry and radiobiology had a place in the factory; however, it took years for so questionable a concept to become believable.
As a result, all three went to the institute in Sungul as soon as it was
founded. They would be able to work in their own professional fields
there.

Timofeyev-Ressovsky[1] came to Sungul. His fate deserves a special
presentation because it is representative of events in the postwar years
when Stalin was the dictator. Timofeyev was a Soviet citizen. In the
1920s he was invited by the brain physiologist Vogt to the Kaiser Wilhelm Institute for Brain Research in Berlin. Vogt incidentally had studied the brain of Lenin at the request of the Soviet Union. Timofeyev
remained in Berlin until the end of the war without giving up his Soviet
citizenship. His research, particularly that with Delbrueck and Zimmer,
on the influence of radiation on genes, that is, on genetic properties,
gave him an outstanding reputation. The Nazi regime did not bother
him during its years in power. His son, however, was arrested for making contact with Soviet prisoners and was sent to a concentration
camp.[2] Timofeyev believed that he had nothing personal to fear from
the Russians. For this reason, and because of emotional attachment to
his homeland, he decided to remain in Berlin when the Soviet troops
entered the city. He was, however, arrested after a brief period and
given a 10-year prison term. The same fate befell his colleague Tsarapkin, who also worked in Berlin-Buch as a Soviet citizen. Both are mentioned in Solzhenitsyn's *The Archipelago Gulag* as fellow sufferers.

Timofeyev was treated as an ordinary prisoner and had to bear the
severest deprivations. His health was completely undermined. However, leading individuals in the NVD noted that they had in their hands
a distinguished scientist who could be useful in studies of the hazards
related to radiation in the atomic energy program because of his experience in the field of radiation biology. He was located in a work prison,
and a major was sent to release him and Tsarapkin. They were given
much nourishment and brought to Sungul. Unfortunately, Timofeyev
had lost most of his eyesight because of the deprivation. He could see
the outline of individuals but could not read. I learned of this while still
in Elektrostal and bought two large books on vitamins and their action

---

[1]Riehl, Timofeyev-Ressovsky, and Zimmer had cooperated in research on the
biological effects of ionizing radiation before being rejoined in the Soviet
Union. See, for example: N. Riehl, N. V. Timofeyev-Ressovsky, and K. G. Zimmer
in *Die Naturwissenschaften* (**1941**, *42/43*, 625).

[2]The son of Timofeyev-Ressovsky died under mysterious circumstances in a
German concentration camp. He was probably murdered by the Gestapo.

N. V. Timofeyev-Ressovsky, the Russian geneticist who, while retaining Soviet citizenship, remained active in his research laboratory in Berlin during World War II. He was arrested by the Soviet police in 1945 and sent to a Gulag. He was ultimately rescued and allowed to continue to work in Sungul as a political prisoner. He and Riehl were close friends. He was appropriately honored after the downfall of Khruschev and Lysenko starting at the end of 1964. (Courtesy of Zhores Medvedev.)

and learned that a deprivation of one vitamin (nicotinic acid amide, as I recall) can cause detachment of the myelin sheath of the optic nerve and damage the vision. I acquired the vitamin in Moscow and had Zavenyagin send it to Timofeyev, but it was too late. The damage was irreversible.[3]

---

[3]Otto Westphal, a now retired German immunochemist, informs me that part of the damage to Timofeyev-Ressovsky's eyes occurred near the end of the war, while he was still in Germany, when he was served an alcoholic drink contaminated with methyl alcohol (sometimes called wood alcohol). The Westphal family had very close relationships with Timofeyev-Ressovsky. In the meantime, I have been told that Timofeyev-Ressovsky did recover much of his vision.

Timofeyev retained the status of a criminal prisoner; however, he was accommodated very well in Sungul. He obtained a house fully as fine as those planned for the German group. He received a post as head of the biological division of the institute in Sungul, and he was allowed to bring his family from Germany—and all that for a prisoner under punishment! It was especially grotesque for him to find a wreath of flowers put out as a greeting when he first arrived at his home. This incident does not in itself give me any stimulus to request greater humane treatment of prisoners in our country. How humane and chivalrous it would be for our female prisoners to find a bouquet in their cells upon arriving in jail.

Although I knew that I would meet several friendly colleagues in going to Sungul, I wanted to inspect the institute and its associated facilities before agreeing with Zavenyagin's proposal. It would still be feasible at this stage of events to make special requests and conditions. As a result, I undertook a trip to Sungul to see the state of affairs. I also took my eldest daughter and a German girl along to give them an interesting journey in the Urals.

The cheerful reunion with Timofeyev and the other colleagues as well as the special things we experienced at Sungul removed all doubts about taking over the post as head of the institute. There was only one matter on which I wanted additional clarification: the head of the chemical division was S. A. Vosnessensky, who was also a prisoner and had the same privileges as Timofeyev. I did not yet know him and wanted to learn during my visit to Sungul whether he was a trustworthy individual. The initial atmosphere for our conversation was very icy. I pressed him in order to stimulate him to a more open-hearted discussion. He remained, however, very reserved. He did not know me, and the golden star on my breast gave him doubt that he really could be open. (On journeys and official visits, I always wore the star and the medal going with the Stalin Prize because they opened many doorways.) Finally he came free and told me of his misfortunes. He had spent perhaps half a year at a scientific institution in Germany before Hitler took power and had gained a great deal from the stay. At the outbreak of the war in 1939, and after he had returned home, he was taken prisoner by the Russians and given a 10-year sentence for being "a potential member of a fifth column." He said that he had never said or even thought of anything that could not be shouted out loud at Red Square. We became good friends.

Satisfied with the investigation in Sungul, I started back with the girls. It is difficult to travel in the Russian provinces without making an excursion into the field of entomology (the study of insects). Our automobile broke down on the way to Sverdlovsk, and it became necessary to spend half the night in a primitive farmhouse. The girls played cards with the escort, whereas I lay down on a wooden bench and slept. Later the girls had countless flea bites on their legs, but I had none at all. The reason lay not in the fleas' taste for nourishment but was a result of the fact that they could not jump as high as my bench. The recorded height for such a jump is about 35 cm (14 inches). In any event, this was the case before the start of consistent requirements for performance in competitive sports in the Socialist countries.

My family and I moved to Sungul in September 1950. Essentially all members of the group in Elektrostal left to go to other places. Only Ortmann came with me because, as an expert in the field of luminescence he would fit in well at the institute.

The atmosphere at the Sungul Institute was entirely different from that at the Elektrostal uranium factory. The cultivated environment was set by the Soviet and German scientists. Almost all the former were more or less prisoners, or at least exiles, for political reasons, without exception.

There were criminal prisoners within the service staff. There was even a murderer whom we referred to as "Our Murderer."

If it were possible to choose the life course of either a political or criminal prisoner, one would be advised to select the latter. The criminals became fully entitled citizens after being released, whereas the political ones had to continue suffering under a stigma. Some of the political prisoners whom I knew had a number on their identity passes. (I believe it was the number 39.) This meant that they were not allowed to settle in a large city. The scientist–prisoners could get their terms cut to half as a result of good working behavior. In this way, most became free from the status of being prisoners in about half the time of their initial sentence.

In order to get to more cheerful things, I should say that Sungul was situated in a region with a beautiful landscape. The institute, the homes, and all auxiliary buildings were in a beautifully wooded, partly cliff-lined, narrow peninsula a few kilometers long. There were many islands dispersed in the surrounding lake. There was a magnificent view of the Urals on the west. The pretty house in which my family and I lived stood on the lakeshore above a steep rocky cliff. If the feeling of being restricted had been absent, one could have resided there cheerfully. The peninsula was barricaded on the land side by a barbed wire fence, and there were even guard posts with alert guard dogs distributed along the lakeshore. The Germans needed an escort if they wished to leave the area. The criminals could not leave the compound unless they were very ill. As a result, we were able to enjoy noble nature only under very restricted conditions.

The climate there was intensely continental. The very cold winter lasted somewhat longer than we would have liked. The winter temperature frequently dropped below –40 °C (–40 °F). I can never forget a trip

Riehl, his wife Ilse, and youngest daughter on the steps of their home in Sungul in the Urals during their brief summer stay there. (Courtesy of the Riehl family.)

from Sverdlovsk to Sungul in a scarcely insulated jeep when the temperature was –42 °C and a stiff wind was blowing. It took the entire night because of the snow, and we nearly froze in spite of warm clothing. Only when the upper edge of the rising sun appeared, donating little heat but the promise of deliverance, did we feel somewhat warmer. I often wondered about this purely psychological effect. The cold winter is followed by a very brief but exhilarating spring. The summer was brief and predominantly beautiful. The plant world thrived more rapidly and more luxuriously than in Central Europe. The flowers were intensely colored; many plants that we knew only as low weeds shot up two meters high; the wealth of wild strawberries in the woods was enormous.

The activities in Sungul were primarily focused on problems related to radiation chemistry and radiation biology. Let me describe the most prominent. Alongside the development of methods of dosime-

The Riehl's youngest daughter Irene dressed for a typical winter day in Sungul.

try were investigations of the absorption and subsequent release of radionucleotides by various organs (the uptake of radioactive isotopes by organs). We also studied, on a statistical basis, the biological effects induced by the incorporation of nucleotides and by penetrating radiation. In addition, we determined the highest permissible doses allowed as a result of exposure to radiation and to nucleotides.

The general public is too little informed of the great rigor and care that goes into the determination of such safety factors. As a result, much of the concern that laymen express about these topics is quite unjustified. Those concerns arise, of course, from the very real dangers that must be taken into account when dealing with nuclear reactors. The same problems in that area are, indeed, exceedingly serious and must be dealt with by trained experts and not by amateurs and half-professionals who wish to be benefactors of mankind.

At Sungul, both Zimmer and Born were able to work on the same type of problems with which they had been involved in Germany. Zimmer worked in radiation dosimetry, a field in which he gained distinction over many years. Born, in turn, dealt with radiochemistry. Whereas he had worked with very low concentrations of radio nucleotides in Germany, in our Russian institute he was dealing with preparations that had far higher activities. In addition, he acquired a wide variety of radionucleotides, obtained from a factory associated with a nuclear reactor not far from Sungul, that were produced as fission products . Another colleague, A. Katsch, focused primarily on the problem of developing methods to extract radionucleotides that had been incorporated in various organs. This involved the use of chemicals that would form complexes with the radioactive atoms as well as the use of other substances. Members of our German group were very fortunate in being able to work in areas of interest so closely related to their previous ones during this postwar period. It made it possible for all of us to continue our work when we eventually did return to Germany.

Timofeyev-Ressovsky was not able to work in his normal field of research, genetics, at that time because of the unfortunate Lysenko affair that I discuss briefly in Chapter 17. The same was true for all of the very earnest Russian geneticists. As a result, Timofeyev-Ressovsky changed areas, investigating the influence of the emissions of radioactive materials on the growth of useful plants. He was able to return to the field of genetics only after Lysenko's downfall.

In contrast to Timofeyev, his long-time colleague Tsarapkin refused in a hard-necked way to find an alternative program that would fit into the program of the institute. He closed himself off and worked exclusively on theoretical problems in genetics. He was not reprimanded in any way for this, but his posture could cause him to risk forfeiting his right to have his prison sentence cut in half as a result of doing "useful" work. I had a long discussion with Tsarapkin, trying to convince him to do something that was at least peripherally "useful," because a shortening of his period of confinement would be of advantage to his career and to his family. I promised him all the support and other things that he would need to make the transition easier. He thanked me for the sympathy and the offer of help, but he was inclined not to accept a pragmatic stance. He felt that there would come a time when someone would take his work on genetics out of its treasure chest and appreciate its value. Only that was of importance to him. In order not to destroy his will to live, I could not bear to tell him my fear that his work would most probably be removed from his treasure chest by one of the members of the security staff, numbered, tied up, sealed, and put away so "safely" that a professional who could read it and understand it would never see it. I could understand Tsarapkin's position but could not agree with it. There are occasions when it is better to render unto Caesar what is Caesar's. I left his room with a feeling that was a mixture of admiration and pity. After we left Sungul, he was transported to a place somewhere in central Asia where he died soon thereafter.

As a result of being director of the institute, I could personally work on topics of interest to me that were just on the edge of the agenda of the institute. Our library contained an excellent collection of the latest journals, including the foreign ones, so that I could follow more or less contentedly what was happening in related fields. What was missing, however, was personal contact with individuals who were correspondingly informed. In fact, I had first come to know many of my Russian colleagues as a result of participation in international conferences. When I first came to the Soviet Union, the ongoing president of the Soviet Academy of Sciences of the Soviet Union, S. I. Vavilov, who had also worked in the field of luminescence and had written a preface to the Russian translation of my book, had attempted to make contact with me and have me give a lecture. However, even this highly placed individual was refused contact with me on the basis that it threatened security.

The pleasant personal atmosphere that prevailed in Sungul was not the result of my actions as director of the site. It stemmed primarily from those of the local NKVD colonel named Uralets, a warm-hearted and wise individual. Most of his inmates never learned of the steps that he took to make their lives easier—steps that could lead to punishment for him. I knew of many of his actions because he undertook them within my area of authority and I had to provide auxiliary support. Free of all ideological stolidity, he handled everything pragmatically and elastically. He was not of Slavic–Russian ethnic origin; his background was Tatar or Georgian. As far as physical features, he resembled the famous Russian path-finding scientist Przhevalsky. Along with his

Sergei I. Vavilov, president of the Soviet Academy of Sciences, chairing a meeting at the academy. He was a physicist involved in research in radiation-induced luminescence and, as a result, was professionally interested in Riehl's research in that field. (Courtesy of Patty Ratliff and the Russian Pictorial Collection of the Hoover Institution Archives. Photo by B. Velyashev.)

organizational talent, he had another characteristic that is rarely found in Russia. Most true Russians have little interest in the state of their natural surroundings. They frequently neglect their gardens and cemeteries and the surroundings of their homes. This behavior became even worse during the period of rapid industrialization. Uralets, in contrast, spared no effort to preserve the natural surroundings of the settlement while it was being constructed. He fought to preserve each tree when the builders wanted to uproot them out of hand. If I place Colonel Uralets in my list of good Russians, I do so not because he had typical Russian attributes in these respects but because he did not.

In concluding this chapter, I must mention a very sorrowful incident that occurred in the area where we had lived in the Urals many years after we left. It is related to the "Kyshtym catastrophe," an explosion in 1957 or 1958 that contaminated a considerable area with radioactivity.[4] Kyshtym is a railroad stop on the Chelyabinsk–Svedlovsk Line, close to Sungul. The small town of Kasli and the first nuclear site are also nearby. The Soviet authorities kept the incident secret, but there is much testimony to support the conclusion that a disaster did occur. The scientist Z. Medvedev, who now lives in the West, collected all the available material on the subject and published it in a book cited in footnote 4. I can affirm from my own knowledge of some of the facts that everything in it is reliable.

What happened? It was clearly not caused by an operating reactor. Instead it was a less serious event, an explosion or venting of material from a dump containing nuclear refuse that had the effect of hurling radioactive material into the air. It was, however, a serious accident that involved the loss of lives, probably including those of some of the individuals who had worked with us. Because the results of the study of the cause of the accident by Soviet scientists are not publicly available, we must guess at the origin. In as much as the Kyshtym catastrophe was a single event, one is inclined to believe it was related to special circumstances involving local geographic, geological, or climatic conditions. Perhaps the hectic conditions under which the first

---

[4]The Kyshtym disaster occurred on September 29, 1957, and was the result of a thermal explosion of a large underground storage facility for radioactive waste. The cooling system that carried off the heat generated by the radioactive material failed, and the tanks containing that material exploded. The explosion was thermal and chemical, not nuclear, in nature, but the effect was, nevertheless, a major disaster because of the radioactive material in the tanks. The plume rose about one kilometer and contaminated a large area, exposing perhaps 100,000 individuals to serious levels of radiation. See, for example, Steven J. Zaloga's *Target America* (Presidio Press: Novato, CA, 1993, pp 219-224) and Z. A. Medvedev's partial account of the disaster in *A Nuclear Disaster in the Urals* (Vintage Books: New York, 1980).

nuclear reactors were built that I mentioned earlier played an important role. Because an open factual analysis and report concerning the source of dangers associated with nuclear reactors would be valuable for all nations, I appeal to the Soviet Union to provide the most open and detailed account[5] of the accident to the professional community. Perhaps the beginning of "Glasnost" will do some good.

---

[5]An account of recently released studies of the consequences of the Kyshtym disaster and related problems resulting from careless dumping of very large amounts of radioactive waste in the Techa River alongside the Kyshtym plant appears in Michael Balter's "Filtering a River of Cancer Data" [*Science* (*Washington, DC*) **1995**, *267*, 1084].

Chapter 9
The Final Battle over Returning Home

In 1951 our freedom was restricted further. We were now permitted to correspond only with our nearest relatives. I decided that this gave us good cause to begin an unconditional battle to go home.[1]

I wrote two letters to Zavenyagin, sending one but temporarily keeping the other. I said in the first that I had essentially no relatives in Germany and, as a result, desired to write to individuals who were not relatives, particularly to an elderly woman who was caring for the grave of my oldest son.[2] I knew from experience that this letter would probably go unanswered. As a result, the second letter stated, "Since my recent letter containing the request for an exception to the rules has not produced an answer, I conclude that you are no longer interested in us foreign specialists. Therefore, I declare that I am no longer pre-

---

[1] Riehl's request in 1951 to return home came at a very difficult time for his Soviet captors. Although the development of pure fission weapons and the means of delivering them were well in progress, the leaders were racing to push fusion bomb technology, now a matter of high priority, in competition with the United States. Actually, the Soviet interest in such technology had started immediately after World War II, presumably at the suggestion of Y. B. Khariton supported by Y. B. Zeldovich. By 1951, Andre Sakharov, working in close association with colleagues such as Zeldovich, was the counterpart of Edward Teller in the Soviet Union, employing a succession of "ideas" that led to success in stages that culminated in 1955. Successes with earlier designs left no doubt about the course of events, politics aside. Riehl, of course, knew nothing of this. Of note is that by 1952 the Soviet leaders were sufficiently confident of their own prowess that they were willing to give Riehl the assurance that he would be free in two to three years.

[2] Their son died of natural causes and not in combat or as a result of bombing.

Yakob B. Zeldovich, a brilliant physicist and close associate of Andrei
Sakharov. The two had been students together and developed close links
in the exploitation of nuclear energy. Zeldovich is credited with being
one of the first of the Soviet scientists to give serious thought to the
release of energy by nuclear fusion. Like Edward Teller, he began basic
analyses during World War II. The topic was pushed ahead in a practical
way soon after the end of World War II. Sakharov eventually became the
leader but worked in close association with a team in which Zeldovich
was a prominent member. For understandable personal reasons, Zeldovich did not feel free to support Sakharov when the latter began
attacking his government's armament policies in the 1960s. Sakharov
never resented this and delivered a moving eulogy for his old friend
when Zeldovich died in 1987. (Courtesy of Photo-Novosti RIA, The
Embassy of the Russian Federation in Washington, DC.)

pared to work for the Soviet government after July 1, 1952." This "letter of resignation" was thought through carefully. I knew that at this time one should not refuse to obey working in the name of a group, because in that case you would be considered to be the a ringleader and could be subject to a prison sentence. A refusal to work on a purely individual basis was less dangerous. However, when it was made clear to the authorities that I no longer desired to remain in the Soviet Union, in spite of my birth in St. Petersburg and knowledge of the language and in spite of all the honors and money I had received, it would be clear, nevertheless, that I was speaking for the entire German community. The Russians could be expected to view the situation in that light and act accordingly.

Andrei Sakharov in about 1950 when he, working with close colleagues such as Yakob Zeldovich, had taken the leadership role in advancing designs for the Soviet hydrogen bomb. (Courtesy of Photo-Novosti RIA, The Embassy of the Russian Federation in Washington, DC.)

As expected, the first letter was not answered. I sent the second letter after waiting for a month. I had previously informed my German co-workers about my actions, requesting that they keep them quiet and recommending that they hold their distance from me if they wished to do so. None of them, however, did. No letters were ever sent by regular mail but were given to the Head Office of the Institute. My letter was not held up but was dutifully sent on to Zavenyagin. After a time, Uralets said to me that I should not get restless. Zavenyagin was very busy at the moment and would presently have me come to Moscow for a discussion.

I was called to Moscow at the beginning of 1952. Typical of relationships in the Soviet Union at that time, none of the staff in the Sungul unit could go to Moscow of their own free will. This was true even for Uralets; he had to request permission from Moscow. Uralets told me that he had once used some of his private vacation time to visit one of the ministries on official business.

I had no idea what would happen to me in Moscow. I did not think that I would be arrested, but it was possible I would be left to stew in my own juices until I weakened. As a result, I procured as many gold bars as I could from the savings bank to provide for my family for two years. I said that I would stay away two years if that proved necessary.

On my arrival in Moscow, matters proceeded in a very suspicious way. I was not put up in a mansion such as "Yagoda's Dacha" mentioned in Chapter 2 in which I had met the other Germans. Instead I was placed in a sanitorium on the edge of Moscow that was empty in winter. I was told that Yagoda's Dacha was being renovated—an obvious lie. On the way to the sanitorium I stopped at a large tobacco shop and bought enough cigars to last me for half a year. (The chief saleswoman running the shop was more fond of me than anyone else in the Soviet Union because I occasionally made enormous purchases of cigars, allowing her to satisfy and even exceed her sales quota.) In addition to me, there lived in the enormous sanitorium building at that time only the escort, the cook, and a porter. I could take walks in the park of the sanitorium with the escort. Several days went by without anything happening, and I concluded that they actually had decided to let me stew for a while. One day, however, I received news that I had an audience with Zavenyagin the next evening. I did not smoke for the entire day before the meeting because I wanted to be in a sufficiently aggressive mood to negotiate the matter of our release. The proceedings went as follows. I had hardly entered Zavenyagin's office when he opened matters with the words: "You write pretty letters to me!" I answered testily, "Yes, and you answered my first letter in a very kind way!" A long uncommonly labored exchange of words occurred. I found myself in such an excited state that I cannot reconstruct the exact details of the exchange. I recall only one assertion of Zavenyagin because it shocked me and raised my anger still further: "On the whole I find your behavior difficult to understand since you have raised on your own the matter of gaining Soviet citizenship some day." I denied this assertion, but Zavenyagin said, "It is true. I witnessed it!" (It occurred to me somewhat later that in the first encounter with Beria I had made an incidental remark that could have been misinterpreted in this way.) At this moment our discussion was interrupted by a telephone call. I experienced a very intense few minutes while Zavenyagin was occupied with the call, because it appeared that I would clearly be placed under very great pressure. Zavenyagin, however, did not return to this dangerous theme when he had completed the call. We were both somewhat exhausted, and the conversation continued more quietly. He said that I would be given complete freedom as long as I stayed in the Soviet Union and that I could select the nature and place of my activities as I chose. Because I retained my position of refusal, he said finally, "Stay in Moscow a few days longer. Consider all of this further and then come back to me."

I returned to Zavenyagin a few days later. On this occasion I did not stop smoking, because I had the impression that the battle over departure had really worked against me. I had hardly entered Zavenyagin's room when he said, "I can see from your face that you have not changed you mind." There followed a long, quiet discussion. I attempted in a most friendly manner and means to explain clearly to him why we wished to leave the country. That was not easy, indeed, it was almost impossible. At least at that time, the fundamental value of personal freedom in a liberal state was inconceivable to the people in the Soviet Union. As a result, I provided the following, more basic, reason: "You must well understand that every person eventually likes to return home." This approach appealed to him most. It was clear that as bearers of important secrets we could not return to Germany immediately. As a result, the discussion drifted to the question of what should be done with us during this period of "quarantine." I said that, if necessary, we would accept a simpler mode of life, even living in a hut. Zavenyagin objected, however, implying that we would find it very hard to endure a life of that kind. It would be necessary to find a good living accommodation and a meaningful occupation. Finally he said that he would discuss the matter with Beria and give me information in due course.

I was called back to Moscow by Zavenyagin some months later. He informed me about the plans that had been made concerning the quarantine period for the Germans who had participated in the atomic project. The quarantine time would be for two or three years. The two centers at Sukhumi on the Black Sea would be transformed so that they

Architect's sketch of the home near Moscow that was awarded to the Riehl family. The family never used it because Riehl feared that accepting such gifts would produce additional entanglements. (Courtesy of the Riehl family.)

would not deal with secret work, and all of the Germans would be concentrated there for the quarantine period. Hertz and Thiessen would leave that site and be located in Moscow for an additional period of secret work. In mentioning here "all of the Germans who had participated in the atomic project," I do not include those who were engaged only temporarily in the various atomic projects and were later removed because no use was found for them or because they refused to work. Most of them were previous prisoners of war. Many of them thought that they would return home sooner if they refused to work. Actually, they were not returned much earlier than we were and were forced to live under essentially inhuman conditions in the interim. I know of their fate only from hearsay. At the end of this book I have cited a book by H. and E. Barwich that deals with this matter.

At the conclusion of this visit with Zavenyagin, he requested me to visit, at least briefly in the few hours left to me, the recently completed house in Moscow that he had provided for me as a gift. I begged off. Zavenyagin said with a trace of bitterness in his voice, "He refuses each time to accept the beautiful house that we have offered him as a gift."

Chapter 10

On the Caucasian Coast of the Black Sea

In the early autumn of 1952, we journeyed to the vicinity of Sukhumi on the Black Sea (see picture on page 84). Located there were two compounds in which German specialists had been living and working since 1945. They were at Agudseri and Sinop. Before the work being carried out was shifted to the unclassified (open) level, Hertz and his group were at the former and von Ardenne and his group were at the latter. With the change, Hertz went to Moscow, whereas his colleagues stayed behind to wait out the period of quarantine. The Germans in von Ardenne's group also stayed behind and were distributed between the two sites for the same waiting period.

My family and I moved into the villa in Agudseri that had been occupied by Hertz. It was one that had been attractively designed by the wife of Volmer and it fit in well with the tropical landscape. Unfortunately, the construction was miserable. Among the wartime and criminal prisoners involved in the construction there must have been individuals from all types of callings, but none who had experience at construction. When I was still in the Urals, Zavenyagin asked whether Hertz could move into the house in Moscow that had been given to me. I readily agreed.

The Caucasian coast of the Black Sea is subtropical because it is shielded from the north wind by the mountains. The vegetation is visually splendid. We had excellent mandarins, figs, and other fruits in our garden. We had so many grapes on our vines that we could harvest only a fraction. The entire coast has always been a favorite vacation area. Our institute previously accommodated a sanitorium. This sanitorium (as well as two others in Gulripshi) were founded at the turn of the century by the philanthropic millionaire N. N. Smelski who desired to provide clinics for individuals who were not well-off, particularly those in the early stages of tuberculosis. The sanitoria he created and supported were exceedingly comfortable as well as inexpensive. The sanitorium building is situated in an enormous park with exotic trees and shrubs. Smelski had imported an entire shipload of trees from Japan. The present-day, average-level Soviet people do not go there now.

I obtained this information about the history of the sanitorium partly from elderly inhabitants and partly from a highly detailed travel book published in 1911 that I had acquired in Germany. Compared to it, the present-day politically colored travel books obtained in the Soviet Union are ridiculous. Essentially everything historical that relates to pre-Soviet times or that is in conflict with Soviet ideology has been filtered out. Thus, Russian history is continually shrinking for the reader of Soviet books. Even events that were essential for the Soviet epoch are absent. Trotsky as well as Beria are hidden, and Stalin is well on the road to being eliminated. (The best chance for

The home in Agudseri on the Black Sea occupied by the Riehl family during their final period in the Soviet Union (1952–1955). It had been designed by the wife of Max Volmer and was previously occupied by Gustav Hertz. (Sketching by Riehl.)

ending up in the Soviet Valhalla is to die naturally by chance, as was the case for Lenin, or to commit suicide as Kirov did in 1934.) The names of enterprising individuals who had outstanding achievements to their credit, such as the great pioneers, are muted, as well as the unusually important history of the Stroganovs who opened the route across the Urals and occupied Siberia. Such individuals are hardly mentioned, even in books dealing with science.

The ever-popular and beloved classical Russian novels that have to do with the soulful rich idlers or the more or less dissolute outsider nobles provide no indication of the much wider, more wholesome and effective layer of productive Russians and foreigners, of the enlightened landowners, of the middle class farmers, or of the merchants and numerous capable officials, such as ministers Stolypin and Witte. If these truly capable individuals did not exist, but only the proletariat and tea-drinking, starry eyed intellectuals, how was the enormous network of Russian railroads developed, or the once flourishing trade, or the previous export of excess grain? How about the fostering of art and science? How about the existing and organically still-growing industries? From whence came the splendid buildings, museums, and theaters that even today outshine everything else the land has to offer? All of this occurred during the 200 years since the time of Peter the Great—and without Lenin!

Let me leave this excursion into Russian history and return to that of my group of German colleagues sitting on the shores of the Black Sea. We did not have any rigid program to guide our activities. My own interests naturally turned to familiar issues in solid- state physics and anything that had to do with chemistry. The group of scientists who were previously under the leadership of Hertz and von Ardenne concerned themselves primarily with problems linked to mass spectroscopy and electronics. A very congenial atmosphere evolved, both at work and in our social relationships. Additional relief from stress resulted from the practice, now feasible, of taking automobile trips into the numerous interesting regions of the Caucasus, but always accompanied by an escort. Some politically interesting incidents that occurred during this period were described in earlier chapters.

Our period of quarantine came to an end in two and a half years, and it was time to return to Germany. The period of between two and three years that Zavenyagin had promised was honored. He called me just before we left and asked me what should be done with my house. I said that it should be converted into a children's home or used as an accommodation for foreign scientists. He said, however, that that was not legally possible, because the house was regarded as my private

Riehl and his daughter Ingeborg as guests of a native family during a visit to the High Caucasus. The family belongs to the Svanian ethnic group. The Caucasus mountain area has one of the most varied populations on the planet.

Riehl on a ride during an outing through a portion of the Caucasus with his inevitable cigar. (Courtesy of the Riehl family.)

property.[1] We finally agreed that the house should be sold. I signed an authorization for sale before leaving the Soviet Union.

I had looked forward to an idyllic return journey through the whole of Russia by train. I had hoped to spend the journey looking out of the window as I took leave of my native land. The idyll was, however, never realized. For one thing, I started the trip very drunk. An especially pleasant Georgian family decided to throw a party to bid us farewell and passed around the table again and again an enormous horn filled with wine. Even worse, however, I developed a horrible tooth-

---

[1]"Ownership" of private property was well-recognized by the Soviet authorities in such cases. It could, however, be confiscated by the government if it so chose.

ache. When we reached Moscow, I had to spend the interval before leaving visiting a dentist instead of taking a relaxed tour of the city that we knew so well.

At the border station of Brest-Litovsk, Colonel Kusnetsov, who accompanied us, said in parting, "The doors of the Soviet Union will always be open for you." It pained me then and continues to do so that the door when open is more likely to be associated with an invitation to leave than to enter. Does it have to be that way? Will it always remain so?

Our train rolled into the station at Frankfurt on Oder in East Germany on April 4. The platform was as if swept clear and was sealed off by the police. One could well imagine that a train bearing lepers had arrived. The arrangements for greeting our arrival had, indeed, not yet

In the study of the family home in Sukhumi. The struggle for freedom is nearly at an end. (Courtesy of the Riehl family.)

been developed. We were experiencing the chilling winds of the German Democratic Republic for the first time.[2]

Because this book is intended to deal only with the time we spent in the Soviet Union, I shall not describe events that occurred during the period that we spent in the Eastern Zone, although many experiences, such as a meeting with Ulbricht, would be worth relating.[3] I do, however, as a result of my knowledge of the Soviet Union, want to raise the issue as to whether the situation in East Germany under the control of the Soviets had to be as unpleasant as it was. I recognize that the domination of East Germany by the Soviet Union, although bitter, was unavoidably necessary. After such a terrible war the Soviet Union did not want to abandon its politico-military bulwark against the West. The trauma of the war even had a subliminal effect on our Western friends. Many feared German bombs less than the threat of Soviet ones. (*De gustibus non est disputandum.*) My concern here is with the question of whether it was really necessary for the leaders of the German Democratic Republic to be so completely servile toward the Soviet Union. I have a definite feeling of competence concerning this issue in keeping with my own experience. The Russians tend to react with great reserve and hardly unconcealed repugnance to the crude currying of favor. Not only I but many other Germans in the Soviet Union found that they fared much better when they preserved a dignified posture than when they behaved in an obsequious way. Where and why did this display of servility of the East Germans originate? The Russians are particularly unmoved by such action, for they have well-developed instincts for distinguishing between real and feigned displays of friendship. Is it the German desire for perfection that drives them to become more like the Soviets in Russia than those in Russia are, or did the Russians really force the Germans who were under their subordination to adopt this posture? If that were indeed the case, it would be necessary for me to cross out the chapter on the good Russians.

---

[2]The brilliant success of the D-Day Invasion, although costly in human terms, spared Western Europe from the possibility of an alternative fate. Poland, Czechoslovakia, and Hungary were not as fortunate. If the Normandy invasion had failed, the U.S. nuclear threat, emerging as it did in mid-1945, might have held the Soviet leaders in check at that time, as it did during the years of the Cold War. Soviet influence would, however, have been far stronger in Western Europe. This would also have been the case had the war in the Pacific continued for a year or more longer.

[3]Riehl's daughter, Ingeborg Hahne, informs me that the meeting with Walter Ulbricht, chairman of the Governing Council of the Communist Party and head of the German Democratic Republic, took place at a banquet and did not involve a very private discussion.

Riehl on an excursion boat with other scientists during an international conference. The photograph was probably taken in the 1960s or 1970s. The conference was probably devoted to the field of luminescence, one of his lifelong scientific interests. (Courtesy of the Emilio Segrè Visual Archives of the American Institute of Physics.)

Abandoning tempting offers and giving up much of material value, including two lovely houses, I went to the West in June 1955.[4] After 22 years, I could finally speak, do, and go as I wished.

In concluding this biographical part of the book, I would like to mention the psychological effect that others who returned from Eastern Europe may have experienced too. Upon becoming a university professor in a quiet eddy, and having the most friendly relations with col-

---

[4]Had Riehl decided to remain in East Germany, he could have retained all the wealth he acquired in the Soviet Union, with some appropriate form of exchange rate between rubles and East German currency. Retaining such funds was probably an inducement for other Germans to remain in the communist Eastern Zone, but Riehl preferred freedom above all else for his family and was prepared to start anew in West Germany at the age of 54.

Riehl relaxing in his garden near Munich in the latter part of the 1980s.
(Courtesy of the Riehl family.)

leagues, I nevertheless felt that something was missing. Then one day I realized what it was: I missed the danger, the threats, and the struggle. At the very moment I became conscious of this, the feeling promptly vanished. Nothing further stood in the way of becoming a contented citizen of the Republic.

(Added note provided by Ingeborg Hahne:)

While some members of the German group leaving the Soviet Union in 1955 remained on the train during the stop at Frankfurt am Oder to continue on further west, the Riehl family left the train for rest and refreshments after a long journey. Their arrival was known to the East German authorities, and the family was taken to a hotel in Leipzig. When it became apparent that the East German government officials would prefer that they remain in Leipzig or more generally in the Eastern Zone, Riehl, using the full weight of the standing he had achieved in the Soviet Union, insisted that the family be taken to East Berlin, where their original home and work place had been. They were well received there by the East German Academy of Sciences and provided with temporary living quarters. They remained in East Berlin for four weeks during which Riehl used some of the converted Soviet money he had received, which could not be taken out of East Germany, to purchase and furnish a home in East Berlin. The furniture available on the market was of excellent quality. After all this was completed, he and his family packed their belongings in suitcases and quietly crossed to West Berlin and freedom. The construction of the Berlin Wall was still six years off. In the meantime, Riehl had made behind-the-scenes arrangements to have the furniture shipped to West Germany for personal use there.

The home in East Berlin was eventually occupied by others, but the Riehl family succeeded in retaining title to it and finally regained possession in principle following the reunion of East and West Germany in 1990. It and the furniture represented in a sense the more enduring material compensation for 10 years of servitude in the Soviet Union.

Chapter 11
Secrecy: Pursued in Deadly Earnest; Often Ridiculous

The intense measures taken to retain secrecy made our lives exceedingly difficult. The expert Germans and their families could only leave the barbed wire enclosures in which they resided with escorts. My group fared somewhat better because the houses were located in the middle of Elektrostal so that we were able to wander freely within the area without the need for escorts. Actually, we found toward the end of our stay there that this degree of freedom was actually illegal. It was revealed to me at that time that an area of about 200 by 100 meters (600 by 300 feet) surrounding our house should have been encircled with barbed wire and that we should have been accompanied by an escort whenever we left that enclosure. I thought about the deviation from this rule and finally came to an explanation. It was clear that if this had been insisted upon, the regulation would have been at least as horribly hard on the Russians as on us. There was an unbroken agreement between the Russians and me not to make an issue of this rule until we had left Elektrostal.

At about that time, I received an old but still valid order signed by Beria that I had not previously seen. It came into my hands as a result of the conscientious faithfulness of a Soviet woman secretary and described in detail the way in which the German experts were to be treated. The order was frightening. It stated that we should live in a closed area, excluded from the remaining village, and that a food store and movie theater would be set up especially for us. The order also stated that the escort should never be farther than one and a half meters (about six feet) away from the individual being escorted. The idiotic order was never put into effect in this form. Someone had dug it up just before we left and made an issue of it. It was, however, circumvented by delaying tactics.

The overshadowing of the future weighed heavily upon us. I remember how my escort lost track of me for five minutes. It may sound ridiculous, but I radiantly cherished the few minutes of freedom.

The requirements of secrecy were often ridiculous and troublesome, notably in personal areas, but also at work. Every written piece of paper was to be considered secret, even if the chemical formula of water or a poem of Pushkin was written on it. One one occasion, when the secrecy requirements were increased, I went to Zavenyagin and said, "Abram Pavlovitch, your secrecy requirements are strangling us." "Us, too," was the laconic answer. All of the wind was taken out of my sails because his answer represented the truth. Zavenyagin was helpless in the face of those who dictated security conditions. We were as joyful as children when one of the security requirements was dropped or when we were able to slightly countermand the security forces. Unfortunately, I can recall only two such happy occasions.

We had obtained the first uranium production, and the product was stored in an existing building in the area which had been further upgraded. The upgrading consisted in mounting two concentric, encircling barbed wire fences in order that the construction staff, who were exclusively criminal prisoners, could not wander off. The prisoners employed were primarily Soviet soldiers who had been captured by the Germans and returned. On coming home they were not received with flowers and folk dances. Instead, they were imprisoned for a few years, charged with being cowards in the face of the enemy. After the upgrading was complete, the small unit was placed under heavy guard. Two guards were posted at the two entrances, which were separated by about two meters (seven feet).

Deputy Atomic Minister V. S. Emelianov, who had previously been a professor of metallurgy, arrived at the area on business. Because of his command of languages, he spent a great deal of time among the Western nations after 1955, dealing with matters of importance for the Soviet Atomic Energy program. I met him twice at meetings in Geneva in that later period. Getting back to Elektrostal, Emelianov displayed his universal pass upon entering the facility containing the uranium. The first soldier allowed him to pass; however, the second, at the door to the building, questioned his pass and would not allow him to enter. Because of this, the first guard got cold feet and did not allow him to leave. Emelianov was trapped between the two soldiers in an area of about one and a half square meters (about 17 square feet). He flushed red and began bellowing in anger. The sense of gloating I experienced from this situation as a result of the soul-searing frustration I had suffered because of security measures knew no bounds. I broke into noisy gloating laughter. This increased Emelianov's embarrassment. I should add that Emelianov, who was not himself fond of security and with whom I previously had good relationships, was hardly the right person at whom to direct my triumphant laughter; however, there was no better choice available to me. One must celebrate as events make possible. It took over a half hour before Emelianov was finally released.

A second, more amusing, event occurred when I was visiting the Atomic Ministry in Moscow on pressing business with our chief engineer, Golovanov. He had a high-level permanent pass for the Ministry, whereas I had to call upon a minister or his deputy in order to have one

Vassili S. Emelianov, the scientist who worked closely with Minister Vannikov (see photograph on page 98) in guiding Soviet industrial production. Emelianov was a fluent linguist and eventually became a principal representative of Soviet science at many international conferences. (From the Archives of the Russian Academy of Sciences through the courtesy of Zhores I. Alferov.)

sent down to me. Unfortunately, no deputy was available in the building. We decided to resort to a trick of impersonation. I went slowly by the guard, carelessly pointing my head toward Golovanov who was following me, and said in a paternal voice, "He is along with me." Golovanov displayed his fine pass carelessly. The soldier must have thought that I was a particularly highly placed individual who did not need to show a pass, and he allowed us to go through. The second posted soldier was so impressed by the scene that he allowed us to pass and saluted.

Several hours later, however, we faced the problem of getting out of the building, which normally required having our entrance passes appropriately signed. It would hardly have been possible to play the same trick, particularly because the guards had been changed. In the meantime, a deputy minister who could provide entrance or exit passes had returned. Taking the bull by the horns, I informed him brightly, as if looking for approval and in an obsequious manner, just how I had obtained admission to his sanctuary without a pass. He signed an exit pass with a bittersweet air, thus permitting me to leave the building. This individual who, as I recall, had the name Meshik, was later advanced to the position of minister of the interior of the Ukraine and was shot at the same time that Beria fell. One can conclude from this that in a struggle toward higher power, Beria had placed his faithful cohorts in the most politically influential positions in the country.

I would like to relate two other special performances of the noble guardians of security that bordered on the farcical. When we were in Sungul, my daughter desired to go to a school in a neighorhood village. She obtained exact instructions from the surveillance office on what we should tell the head of the middle school. These instructions were ridiculous from the start because the people in the village knew, through relatives and friends, who worked on the site and a great deal about us. All were informed except the director of security. Concerning the question of the occupation of her father, she was requested to say that I was a "medical doctor." One day a middle school teacher approached my daughter and said that because I was the recipient of a Stalin Prize, I should be a very good doctor. Would I treat her mother? In her situation, my daughter said that I was a dentist. It remains unclear to me to this day just what service one would have had to perform on Stalin's teeth to receive a Stalin Prize.

The second example of the lack of imagination and the simple-mindedness of the security authorities is even more grotesque. About a half year before we went home, members of the security staff came to me and said, "The Germans (including us) should provide the staff with all copies of birth certificates and school report cards of their children

as well as other documents that they obtained in the Soviet Union."
They wished to reissue the documents with false names of the places
where the documents had originated so that the locations of the secret
sites where we had been working would not become known to the
Westerners! The documents were collected, and we received replace-
ments, all officially falsified, some time later. All the children of the German experts were born in Moscow and went to school there. Unfortunately, the authorities gave the children the same grades that they had
had on the previous cards. There apparently was no "Good Russian"
among the counterfeiters.

Chapter 12

Management of Plans

The many disadvantages and the few advantages of planned management known to me will be discussed, primarily for the benefit of experts. For a lack of competence, I would like to withhold judgment on these issues because it is not clear to me to what extent a factual, non-ideologically supported, trend in economic planning is desirable or necessary.

Both planning and strong directorship are clearly necessary in the case of important military and urgent political requirements, such as the atomic bomb project, particularly when the policies involve great sacrifice on the part of the people. One may recall Hitler's slogan "Guns instead of butter!" However, the unrestricted private enterprise system worked in an uncommonly lively and balanced way in Hitler's area of authority and did so even more so for the Western Allies. In the Soviet Union, however, the industrial sectors were 100% under the forced control of procedures based on economic planning. The evil effect of this process upon the consumer industry is well-known. However, the limitations of planning methods produced continuously adverse disturbances, even in the production of uranium. The following two examples give indications of the consequences of the system.

Toward the end of 1945, we had produced a significant but still modest amount of useful uranium, and I believed the production would then be handed over to the user. That did not happen for the following reasons. To help us achieve, and even over-achieve, the planned quota, we were encouraged by the possibility of obtaining a premium reward for our product that would increase steeply in accordance with the level of overproduction. As a result, it was to our advantage to hold back the available product, which was below the quota, and add it to the production of the next quarter because if matters went as we expected, we would then exceed our quota for that period. I learned that this practice is very widely used in Soviet industry. It follows that the apparently wise premium system sometimes retarded rather than accelerated the release of product. One notes how difficult it is to determine the realities that will result from decisions made at the conference table. My example provides a not uncommon result associated with a planned economy.

One afternoon at the very outset of our activities in Elektrostal, I was told that we should produce a complete list of all the chemicals and apparatus we would need by nine o'clock the next morning; otherwise, our requirements would not be included in the state plan for the following year. As a result, we were careful to obtain sufficient paper and writing implements before quitting time in order to achieve this goal. The entire German group worked all night at a feverish pace, dividing responsibilities in the preparation of the lists of desired materials. The resulting catalog was between one and two centimeters (roughly two thirds of an inch) thick. It represented the pinnacle of German efficiency. Precisely at nine o'clock the next morning, I proudly presented the catalog. Later I learned that our concern was entirely unwarranted. Our needs were inadequately and sporadically satisfied both before and after the exercise. We frequently had to do with improvisations.

The wish list did, however, have one consequence, and indeed, an irrational one. Within the list of chemicals, we had asked for about one or two kilograms of potassium permanganate for laboratory use. Two or three years later, a member of the procurement division came to me and said that the potassium permanganate we had requested was available. I was pleased and asked him to bring it up to me. He thought that it would not be possible for him to do so. I asked him for the reason. He thought there would be too much to carry. I asked him what the quantity was. He responded three wagon loads. It is not clear in what manner two kilos of potassium permanganate had been transformed into three wagon loads. The path of our requests was long and complicated. They first went to our ministry and then, after examination and checking, went to the National Planning Committee. From there, after further examination and checking, they went to the Ministry of the Chemical Industry and then finally to the chemical factory. It is hardly likely that what occurred in this case could ever be determined because no one had any interest and there would always be the fear of what the investigation would reveal.

The potassium permanganate probably still rests somewhere on the enormous grounds of our factory. The affair, however, can be viewed in a more friendly way. When two kilos are ordered and three wagon loads are received, one has experienced a multiplying factor of about 100,000% for the production quota. This provides an astonishing example of the Socialist work ethic.

Chapter 13

Comments on the Technical Training of Soviet Technical Experts

In this chapter, I will intentionally forego my position as a cynical observer of situations in the Soviet Union because we Germans sit in a glass house as far as problems related to technical training are concerned. In describing my experiences with the Soviet training system, I will limit myself to activities at the higher university level of scientific professional areas; that is, I will deal with the traditional universities, the technical universities, and the specialized technical training schools. In the process, I will not avoid taking side glances at our own problems in Germany.

The starting attitude toward professional training in the Soviet Union is completely different from ours. One of the central differences is that German industry grew steadily and organically and was partly based on enterprises involving skilled handiwork. In contrast, the development of industry in the Soviet Union took place very rapidly and intensively as a national enterprise. The factories were developed as integral units "from start." In Germany, the basis was initially provided by a breed of highly qualified, but not academically trained, craftsmen who were exceedingly important for industrial development. The German master craftsmen played a major role in our metallurgical, metal processing, chemical, and optical industries. Such individuals have become much less important today, as one learns through a great deal of public discussion. Complementary to industries based on the crafts are the ones that required university graduates (these industries have been the focus of so much adverse discussion[1] in Germany that I fear for their survival).

The situation in the Soviet Union is quite different. There, national industrialization did not grow out of business-controlled factories employing craftsmen. As a result, there was a much greater need, which was not without consequence, to satisfy the requirements of industry by using graduates from higher educational institutions. To use an analogy derived from building construction, one can say that the Soviets built their industry from the top down. First, this process fit in well with the initial spiritual posture of the Soviet system in which Lenin's declaration of "Learn! Learn! Learn!" always had validity. Second, it derived in part from the spirit of the university system, taken over from Czarist times. Alongside the traditional universities and the technical universities, there were highly specialized technical schools of a type we designate today as "training institutes for the specialists." There was, for example, a school for railroad technology, one for machine construction, one for electrotechnology, and so forth. Their level of quality was often very high; however, they were directed more to the training of specialists in the fields than to general technical education.

My information concerning the activities within the expert crafts during my period of involvement in the Soviet Union went as follows. Industry was continually importing a significantly large number of individuals from the higher educational systems. Indeed, a few of them rose rapidly to very high positions as a result of knowledge and ability. The remainder were spread throughout many levels, and many of them fell under the direction of a highly experienced industrial boss.

If one compares a cross section of the Soviet university graduates of that time with a corresponding German group, one would conclude, on justifiable grounds, that the German group was qualitatively superior. Although we were termed foreign "specialists" in the Soviet Union, it was our deeper and broader training as well as a better knowledge of basic science that gave us special effectiveness in the country at that time. We were often presented with new problems for which we were in no sense specialized. However, thanks to our understanding of the fundamentals of science, we were able to find our way through difficulties fairly rapidly.

In making such a comparison, however, one must be very careful. One cannot conclude from this relative examination that the German

---

[1]Here Riehl is probably referring to the influence of the Green Party and related groups that attempt to place obstacles in the path of technically advanced forms of industry, frequently with success, as in the case of industries related to nuclear power and recombinant DNA research.

universities were necessarily better. Such a conclusion would be based on a comparison of two things that are not really comparable. On the Soviet side, most of the individuals were graduates of the special training schools, whereas those in the German side of our group were products of the traditional universities or technical universities. It is obvious that the level of most of the Soviet universities is as high as ours. Of the Russian technical universities (polytechnical universities), the Leningrad Polytechnic Institute is particularly distinguished, being at a truly world-class level. When I previously mentioned how well the German university education system served the work of the Germans in the Soviet Union, one should ascribe it primarily to the result of our knowledge of fundamental science in specific areas. It is not at all possible to say that the type of education received in the specialized technical schools is inferior in its own way.

It is well-known that instruction in the arts and humanities is dominated by the accepted ideological principles in the Soviet Union. However, the organizational methods used to achieve educational goals of all types are highly pragmatic. This pragmatism is particularly strong for education in science and technology. The Germans who lean toward more democracy in society speak a great deal of "the right to education." The Soviet system is completely free of such sentiments. There, one speaks rather of a "duty toward education," focusing on the interests of the state. The students in the Soviet Union receive a modest salary. That is a necessity today and is required, indeed, even in Germany. Our government pays many students a helpful stipend. The salary in the Soviet Union is staggered, however, depending on performance in school. In other words, the salary is not thought of as a natural benefit of the social system but as a means of encouraging industrious scholarship. Continuing control of the response of the individual to the educational process is facilitated by certain procedures in the educational system. Those procedures, going back to Czarist times, do not depend on a broad examination in a number of topics after the student has been at the institution for many semesters, as is the case in Germany. Rather, an individual undergoes examinations continuously in given topics throughout the educational period. We have what is approximately the same system when testing students at the end of a laboratory course or with the use of a written examination centered on some special lessons. Such practices do not run the risk of undermining the university system by "schooling" it. Rather, it protects the students from pushing aside subjects that do not interest them and from crowding together the study of everything they have learned in a brief period before a comprehensive examination.

The matter of requiring detailed examinations along the way holds only for the early and mid-stage of the higher educational process in the Soviet Union. In the last stages, involving a specialized diploma or a doctoral degree, there is no deviation from the best principles of academic freedom. (The return to such principles has been true for a long time even in the much regimented Soviet Union.) In the later stages, the educational process shifts completely to research, small seminars, and discussions with both fellow students and young postdoctoral staff.

The current practice in Germany of evaluating the level of education of postdoctoral fellows on the basis of the number of courses they have taken clearly makes no sense. (This, now common, bureaucratic policy threatens to strangle the effectiveness of the universities in Germany.) The number of cases in which giving freedom to science has produced idleness is very small. The number of cases in which it has had great consequences are legion.

The procedures developed to educate technical specialists for the atomic energy program followed a much more rugged regime than in other areas of higher education. The students who were singled out for this route during their last semester received special privileges but were also duty bound, after completing their study, to spend three years in an atomic energy center without mentioning the place and nature of their work. I recall two young women, radiochemists, who chose this path at the University of Leningrad. As they rode through the primeval forests of the Urals on the way to us at Sungul, they had no idea where they would land. They were overjoyed as their buses entered the grounds of our institute to find that they were not ending up in a factory but in a scientific institution.

I feel motivated to relate one more piece of practical knowledge gained in the Soviet Union that was of special interest to us. On supposedly rational grounds, there was in the ministries, as well as among university professors, a tendency to create broad-visioned but small professorships and institutes in fields of narrow specialization. Our own experience in the Soviet Union militated against such a tendency because we had such a broad charter at Sungul. During the development of the atomic energy project, however, these small institutes proved to be of great value to us, even though some of them had blossomed in secret for many years. We could frequently make use of the specialities contained in these germ cells and thereby save much time and effort. Fortunately, they had not been allowed to decay over time. I recall, for example, the rapid and effective help I received from an institute focused on vacuum technology, from one focused on photography, and from many others. If such germ cells were once sacrificed for the sake of achieving some spectacular goal, the effect would be irreversible and hence highly regrettable.

Chapter 14

The Condition and Behavior
of the German "Specialists"

From direct observation, I knew only the circumstances related to the German specialists who were involved in the atomic energy group. There were, however, many more working in the Soviet Union (see footnote 16 on page 32 and the related text starting on page 31).[1] I would guess that there were at least 10 groups. I spoke to many of them after returning home and traded experiences. Everything we learned indicated that our group was, in many ways, treated substantially better than the others. The term "Golden Cage" can truthfully be applied only to our group and not to the others.

Nevertheless, the word cage is valid for all of the German workers because none were allowed to leave their site except in the company of a human watchdog. However, the material conditions of the Germans were very different from group to group. The material from which the cages were made ranged from gold to rusty iron.

I mentioned in Chapter 11 the need to have an escort in order to leave the area. It should be noted that the number of available escorts was not so large that every German had one available at any time. This scarcity created an additional limitation on the ability to move about. I had been informed that, as a group leader, my family and I

---

[1]One of the groups was devoted to the development of large-scale rockets and rocket motors. An account of this work appears in Steven J. Zaloga's *Target America* (see footnote 1 in the introductory material). The work of some of the individuals involved in the rocket program is described in the *Nova* documentary (see footnote 2 in the same part). It is interesting to observe that the German rocket team insisted on withdrawing from the program when it realized that developments had advanced to a stage where the rockets under design could deliver nuclear bombs to Western Europe.

would have three escorts available, two for the day and one for the night. That was, however, a purely theoretical regulation. I often had personal difficulty in obtaining an escort. Moreover, I never required the use of an escort at night. It is also not clear to me what use could be made of this privilege. It was never explained to me just what type of night-time escapade was presupposed. Also, nothing was said about the degree of involvement of the escort in such an escapade.

As mentioned in Chapter 11, it was exceedingly tiresome to have a shadow with you. That was not true on only one occasion. One time when my automobile was required to stop by a militiaman, I told him that my escort was my adjutant and I must confess that my self-image was elevated for a brief moment. Actually, I never was in the military. My military ambition was limited to hearing the beat of a Prussian military marching band; and I never once had an adjutant.

Returning to the serious side of life, let me emphasize that the personal relationships between the Germans and the Russian people were good without exception for all of the groups known to me. We were never invited into the home of Soviet families, in spite of the well-known hospitality of the Russians. Perhaps appropriate directions had been given to Soviet families from above in order to discourage this practice. The restriction would stem less from political grounds than from grounds of preserving secrecy.

What must be stressed most is that we never experienced any direct expressions of hate toward us Germans, or even felt it. Teenagers sometimes made minor derogatory remarks in our early days in Elektrostal but were quickly called to order by older people. We know, of course, from many Soviet friends that their families had made great sacrifices in connection with the war. I also stopped in areas that had been occupied by the Germans for a short time. Even then, I encountered no hate. I emphasize specifically that these areas were occupied only for a short time; clearly the behavior of the troops who were under discipline cannot be compared to that of the "golden pheasants" and party big shots. Although the correct and even friendly behavior of the Russian people toward us was understandably ordered to a degree, I believe that the actual source was much deeper in the mentality of the people. I emphasize this explicitly because the attitudes seemed to be conciliatory and revives in one a sense of confidence in humanity.

We could often ascertain the high regard still felt in the Soviet Union for the expert competence of the Germans. I insert "still felt" here because in the main everything American is regarded in the Soviet Union as the model of scientific and technical achievement. This is particularly true for the young Soviet generation, which has little knowledge of the previous merits of German science and technology. In old Russia, German competence in technical matters was illustrated by the widely voiced expression, "the Germans invented the monkeys." I recall a controversial discussion with Zavenyagin shortly after we had succeeded in carrying through our main program on uranium in Elektrostal. He wanted me to undertake the process of retrieving uranium from the slates in the Baltic. It was known that both the Estonian and the Swedish slates contained nominal quantities of uranium, but to my knowledge the Estonian variety had less. I strictly refused to take on a new large and difficult program. To support this refusal further, on the basis of what might be called facts, I said that what he offered was not a problem for physicists or chemists, but for earth-processing experts, a field we did not know much about. "You cannot say that," responded Zavenyagin. "The Germans invented the monkeys."

As much as they wanted us to become involved in expert matters, they were never willing to allow us to take over organizational or other areas that were not purely technical or scientific. When I once offered Zavenyagin some advice in relation to a nontechnical matter, he answered briskly, "I don't need your advice." This unfriendly response to my well-meant comments rankled me then and continues to do so even now. To gain back my peaceful spirit, I will relate in Chapter 15 another example of a case in which advice elicited a Soviet response.

The German group in the atomic center had, for the most part, little reason to complain about material items. We had plenty of food at a time when the German population was starving and the Russian population was in great want. We had permission to acquire food in special stores (limited stores), in which a complete line of specialty items could be obtained at acceptable prices. Only a few Soviet citizens had access to them, indeed, only distinguished Soviet scientists and artists as well as highly placed functionaries and government officials. The supply of fruit was particularly low in both Moscow and the Urals. However, it was far better on the coast of the Black Sea where it could be obtained in the open market. (We could also send packages of fine food back to Germany, as was mentioned in Chapter 7.)

The accommodations we had were also very good. In Elektrostal each family received a finished prefabricated wooden house that had been made in Finland. It had three decent rooms in addition to a bath and a kitchen. There was also a garden with each house. In the Urals, most large families lived in very pretty country cottages made of wood, with five rooms and a glassed-in veranda. Many others lived in flats in a large multistoried stone building. The accommodations on the Black Sea were similar. There, however, my family and I had the special privilege of living in a very large luxuriously planned stone villa. (See photograph on page 142.)

The gap between our standard of living and that of the Russians was enormous. It was notable, however, that we never experienced any envy on the part of the Soviets. Personally, I felt that this difference was painfully embarassing. I recall an evening in Elektrostal when the director of the children's school sought me at home. First he started talking about general affairs, and I did not understand the reason for his visit. After the third glass of vodka, however, he came forth. He asked whether, for the psychological benefit of the Russian children, I would exclude chocolate from the children's mid-morning snack, because the Russian children did not have access to chocolate. I acknowledged the request and followed it.

The legal position of the German experts in Russia was as undefined as you may wish. There was no precedent in Soviet common People's Law to take care of our situation. Our German passports were no longer valid because the government had collapsed. We were not prisoners of war, but we were also not civilian prisoners in the sense considered in previous wars. We were offered employment contracts, but our group had decided not to accept them. It provided us with no advantages and perhaps carried disadvantages. Our position was perhaps comparable to that of those foreigners who, hundreds of years ago, fell under the jurisdiction of a czar or another Eastern potentate for employment and obtained everything with which to live except freedom. We had no means of asserting legal rights. Although two German states were eventually formed, we were citizens of neither one. We learned from the British radio that when an English journalist asked an East German official about the fate of the German scientists who had been sent to the Soviet Union, he replied that there were none!

The presence of the German scientists in the Soviet Union was kept a secret from the Soviet population in general. It followed that we had no proper means of making contact with Soviet scientists, particularly with any who were not working in the same areas that we were. On their part, Soviet scientists hardly had an opportunity to come into contact with us. As you will recall (see Chapter 8), ongoing President S. I. Vavilov of the Soviet Academy of Science (see photograph on page 130), a prominent physicist in the field of luminescence, had wanted me to give a lecture on this field; however, his attempt was forestalled.

In view of our complete lack of legal rights, one had to rely on one's instincts in determining behavior and in making decisions. One could not depend upon previous experience and precedents. One did not know when to be firm and when to yield. Personally, I had, on occasion, obtained good results from a display of anger. However, it is obvious that one will soon exhaust this weapon if it is used too frequently. On occasions I even ran proudly through open doors. In general, however, the better way is to trustingly allow things to come along as they occur. In such an unpredictable situation, it is best to base one's decisions on the counting of one's buttons (the toss of a coin). One then makes the right decision in half of the cases and can be proud of having such a high success rate.

As was mentioned in Chapter 9, a number of German war prisoners were working in the atomic group. Acquiring them was the result of an initiative first taken by von Ardenne. He told us with some enthusiasm that when he was taken prisoner by Beria in June 1945 he requested that German war prisoners who were scientists, technicians, and mechanics be taken out of the prison camps and placed for work at the disposal of the specialist groups. This proposal was accepted. The plan was a good one from a pragmatic and a human viewpoint, but in practice it did not always have the desired humanitarian result. Whereas all other war prisoners were sent back home by 1949, those who worked in our specialist groups had to wait much longer. Because this result was predictable, I had reservations about acquiring war prisoners without providing them with the facts. As a result, no such prisoner was made a part of my group against his will. I was able to achieve this end as a result of my familiarity with the Russian language.

The selection of the prisoners took place in the following way. They were first sorted out in the prison camps and were then brought to Moscow in groups so that the atomic energy group head could seek out appropriate technical experts. I carried on a "bridal selection" of this type on two occasions. A translator was not necessary because I could play that role. The Russian head engineer who accompanied me did not understand German; therefore, I could talk freely to the prisoners. I told those who had suitable backgrounds that joining us would lead to a good occupation and good accommodations as well as the opportunity to bring their wives and children from Germany. However, I explicitly stated that I could not guarantee how and when they would be able to return home. With one exception, the prisoners reacted negatively toward joining. I then explained to the accompanying Soviet official that those who had displayed concern were not technically appropriate for us so that they were excluded from the group. One of the group agreed to join us with full understanding of the problems that would go with the recruitment. Later, I had a second similar meeting with war prisoners.

A much larger group of war prisoners had been recruited for the atomic group in Sukhumi in accordance with von Ardenne's activities. They either went home with us in 1955 or at most one or two years earlier. Many were allowed to send to Germany for their wives or brides. Others remained unmarried, which understandably led to complications. I shall relate an adventurous incident that went from the pathetic to the grotesque. From the professional viewpoint, the war prisoners fit in well for the most part with the already established German workers.

A few, however, became lax in their work, the principal motive being the hope that they would be sent home earlier than the others—a hope that was practically never realized. They were regarded as bearers of secrets to be kept within the Soviet Union as a result of their residence on the site of a secret project.

Occasionally, some of the war prisoners came to believe that a German soldier should not work in the service of an "enemy," a view that was unrealistic and represented a naive sense of loyalty to the bankrupt empire of Hitler. Such a viewpoint could survive only in the type of isolation that many of the prisoners had endured. Those of us who had been in Germany at the disastrous end of the Third Reich could comprehend events better in their entirety. We saw how the regional leaders and the other party big shots left thousands of individuals behind in misfortune while they saved themselves. As a result, we could not comprehend such a high level of naiveté on the part of the German war prisoners. Almost none of us, of all people, were happy about being brought to the Soviet Union but none of us had acquired the memory associated with shouting the slogans of the Nazi "Werewolves" or had been involved in taking physical action to suppress those who opposed the Nazi regime. Perhaps it is good for the younger German readers to note that the overwhelming collapse of Hitler's Reich produced in us older individuals a sense of the collapse of the world. The younger generation cannot comprehend the extent of the catastrophe and of our sense of personal loss. We, however, appreciated it fully. Many elderly individuals who had not been Nazis took their own lives in despair as a result of the complete loss of everything of value that they had worked for during their entire lives. Under the circumstances, however, most individuals thought of survival first of all and then of some possible form of reconstruction, but otherwise of nothing.

We were left completely alone politically. I was invited to a party gathering only once in Elektrostal because questions regarding production were to be discussed and I was needed as an informed individual on that occasion. I heard, among other things, that biased party issues (and not objectivity) are the principal concerns of party officials.

In Sukhumi, shortly before returning home, a few Germans thought they should acquire some knowledge about "Dialectic Materialism" (*Diamat*) before leaving the country. This was the so-called scientific foundation of Marxism–Leninism. They asked the Soviet authorities to organize a course of lectures. For the uninformed reader, dialectic materialism was treated as a form of pseudo-science in the Soviet Union that can be placed at the same level with and compared to the Nazi "racial theories." The best characterization of dialectic materialism is given by the following riddle:

"What is science?"
"It is the process of searching in a dark room for a black cat."
"What is philosophy?"
"That is the process of searching in a dark room for a black cat that is not there."
"And what is dialectic materialism?"
"That is a process of searching in a dark room for a black cat that is not there and suddenly shouting, 'I have it.'"

The Soviet authorities came to me on this occasion and said that they would provide the lecture course only if I attended; otherwise the other Germans would feel compelled to stay away. I answered with an abrupt "No!" There was an embarrassing pause, then the functionaries asked under what conditions I would attend the course. I held my breath and then delivered a long speech in which I gave the conditions: scientific level, the possibility of completely free discussion, objectivity, and so forth. One of the functionaries zealously wrote down my conditions. Finally, someone said to me that all of my conditions would be fulfilled. They had elegantly placed me in a difficult position because it would now be unfair for me to insist that I would not attend. The conditions I requested were in fact followed completely. A postdoctoral scholar from the University of Tbilisi was obtained, and the lecture course, which was well attended by the Germans, took place in a completely open way. I could not get over my defeat, however, and swore revenge. I achieved it at the third gathering. Following a discourse by the lecturer on the evils of capitalism, I said that what he related was all well and good, but it described the state of capitalism 100 years ago. The situation had changed a great deal since the days of the "Robber Barons" so that the starting basis for the Marxist–Leninist teachings was outdated by 100 years. That was too much for him to hear: the progressive doctrine of salvation out of date by 100 years! The course was terminated.

At this point, I would like to insert my personal reaction to the concepts held within the Soviet Union. In many reports concerning a stay in the Soviet Union, one finds troubled comments by the authors regarding what they have seen and experienced. Often this leads to a transformation of the individual from being a friend to an enemy of Soviet socialism or, in some cases, the reverse. I never had cause to undergo such a transformation. I was free of illusions from the start. As a first-hand witness to the October Revolution in 1917 and of the first years of Soviet communism, I knew the devastation communism produced on the living standard. Even in Germany, an interested and experienced observer could recognize the political similarity of Stalinism and Hitlerism. To repeat, I was indeed surprised at the preference given to us atomic specialists relative to the people of the Soviet Union but, other than this, nothing surprised me. I did find, however, to my pleasure that one can discover true humanity everywhere if one looks.

The attitude of my fellow Germans in the specialist group toward matters in nontechnical areas was not always associated with a feeling of immense pride in being part of the country in which we were compelled to work. That, it should be emphasized, was principally a result of the unfamiliar and unnatural conditions under which we lived: the lack of freedom, the continuous surveillance, the restriction to a small area, and the absence of contact with the rest of the world. Ignorance of the language or the lack of familiarity with the human side of the Russians did not of themselves have much of a negative effect on my fellow Germans, although it has long been known that many Germans have a tendency toward small-minded middle-class behavior that does not often match the Russian tendency toward untidiness. More embarrassing to me overall were the many disputes among the Germans themselves, conflicts exacerbated by the need to live together so closely.

In connection with these matters, I experienced a not atypical incident of German behavior in connection with R. Döpel, mentioned in Chapter 2. He came to Russia as a widower and lived the entire time in Yagoda's Dacha in Moscow. He was a passionate hiker and was exceedingly uncomfortable with the lack of freedom to take long walks. One day he escaped in a forbidden manner, namely through a hole he found in the fence surrounding the park belonging to the dacha and succeeded in taking a long walk. (We all knew about and treasured this hole.) While walking in the woods, he discovered an individual who was dead drunk and lying on the ground. It was late in the autumn, and Döpel decided the man could freeze to death. He went to the neighboring tiny village and notified the local militia in very broken Russian about his discovery. What happened was inevitable. Döpel was arrested, locked up, and located only several days later by the branch of the NKVD to which we reported. The drunken Russian was left to lie. We hope, however, that as a genuine Russian, he was able to withstand the cold and still enjoys a glass of vodka.

Now for a word regarding the status and conduct of the young people who grew up within our atomic group. Many of them went to Russian schools and universities and were quite satisfied with the experience. They lived in the same dormitories as the Russians. The living conditions in these quarters were miserable, but the disadvantages were more than compensated for by the fine comradely relationships with their fellow Soviet students. They had essentially no understanding of the longing of their parents to return home. They had little if any memory of Germany. The memories they did possess were of nights of bombing, which did not encourage a desire to return. Those young people who did return to Germany required a considerable time to become acclimated. The crowded circumstances in Western Europe bothered them considerably. Also, many of them felt alienated from the mindless superficiality of the many over-indulged Germans they encountered.[2]

One may ask in retrospect what aspect of the behavior of us Germans in the Soviet Union was wrong and what was right. We shall start with the first. During the initial period we were there, one could often observe a tendency of the Germans to display brashness. That was a residue of the Hitler period that no German in the country could completely avoid at that time (and in part, can still not avoid). The brash approach occurred only occasionally in as much as the "Herrenvolk" (master race) were confined by barbed wire. The opposite behavior, namely groveling was even less effective in producing results. Fortunately very few Germans in our group exhibited brashness. It made little sense to "insist upon your good rights" because that can be effective only when there is a corresponding way of achieving such "good rights." Moreover, we did not have any formally written rights in our captivity, except the rights that go with appealing to humanity and reason. It was easier to call upon these unwritten rights when trying to satisfy requirements.

In the purely human relationships with the Soviet functionaries, colleagues, and co-workers, it proved best to adopt behavior that is useful everywhere: no overwhelming comaraderie, rather a friendly but somewhat distant posture. It is well-known that one gets closer to people when one maintains a distance.

I had a special reason for maintaining a certain distance from the Soviet people with whom I dealt in order not to seem to be part of their society in view of my knowledge of Russian and my early origins in St. Petersburg. The cigar I had in my mouth and the Homburg hat I wore served as a means of greeting without the need for additional display. Such external appearances, humorous they may be, are very useful in their effect and have an influence not only upon others but also upon oneself. They serve as a form of psychic "corset" or armor in times when one is carrying great burdens. Anyone who has been in a concentration camp, in prison, or in a hospital, knows how important it is to maintain your self-esteem at a very high level by shaving regularly

---

[2]On returning to West Germany with her family in 1955, Riehl's older daughter, Ingeborg, found the normal interstudent comradely relationships at German universities far different from those in the Soviet Union where very lively and even exuberant comradeship prevailed. Her sense of being a stranger in her homeland finally abated when she joined an international student hostel group at the University of Munich where she eventually studied.

and maintaining high standards in order to prevent mental and bodily degeneration. This is called "mental hygiene" in English.

In view of the somewhat rapid build-up of Soviet military technology, Westerners often ask how effective the contributions of the Germans were to the immediate postwar development of this form of Soviet industry. It is my opinion that it would be naive to believe that the cooperation of the German specialists actually played any decisive role in the development of the nuclear energy industry and other important areas of technology. In the field of nuclear energy, the Soviets would have achieved their goals one or at most two years later without the Germans. The significant fact, as I mentioned a number of times, was the enormous concentration of all aspects of available science and technology toward these problems. The motivation for such intense efforts in the field of armaments by the Soviet lay in the trauma that Hitler's treacherous attack had had upon the country. One should recall that this attack came not long after Hitler had signed a nonaggression pact with Stalin! The most beautiful claims of love of peace on the German side neither can, nor will, alleviate this trauma. Hitler woke the sleeping dog, and it will not go to sleep again. I recall a statement made to me by Zavenyagin as he attempted to explain why the Soviets had developed the atomic bomb. He said, "Without it we will lose our sovereignty." Of particular help for the Soviet nuclear energy program in the early stages were the steps taken by the United States and other Western countries to provide the Soviets with a great deal of industrial assistance. It is true that the Americans bombed the Oranienburg atomic plant to hinder the Russians from acquiring it. However, commercial interests of Western firms furnished the Soviets with everything else they needed for the development of nuclear technology. As Lenin once said, "The capitalists will sell us the rope to hang them with."

Something similar to this is also valid—*mutatis mutandis*—for us scientists and technologists. The sale of a length of rope is motivated little by political or ideological factors. The same can be said of the conduct and offerings provided by the scientist or the inventor. The motivation for their activity lies in an entirely different plane. One speaks these days of the "responsibility" of the scientist and attempts to investigate the essence of the motivating force that drives the creatively working scientist or inventor. The motivations are curiosity, the desire for new knowledge, toying with nature, instinct for adventure, the joy of handicraft, and an essential sportsmanlike happiness in overcoming obstacles or contradictions that are encountered. Also, in the case of many scientists, there is the aesthetic appeal of discovering a closed intellectual or mathematical system. These driving forces are linked to what might be called a form of "creative restlessness," which is deeply rooted in the essence of the living being and without which biological development might well not be possible.

Just as unfavorable and, indeed, fatal developments such as lethal mutations can occur in biological systems, technical development can lead to unpleasant results. (This can occur even to a servant of God, such as the monk Berthold Schwarz; instead of inventing something beautiful such as a new liqueur, he invented gun powder.)

Should one, in view of the threat of atomic weapons and other emerging technologies, stop or attempt to control the "creative restlessness," the thinking, the searching, the groping, and the investigating? Terrible thought! Moreover, it is too late. We already have in our arsenals enough lethal material to destroy ourselves. The thoroughly difficult and complex problem, which is partly beyond the very rational problem of preventing World War III with or without atomic weapons, is that of directing the potential for aggression into "harmless" areas. This is and remains primarily a matter for politicians. It cannot be delegated to the imaginary "responsible" scientific investigator. It is not the scientist but the professional politicians who have the authority to either unleash or prevent wars.

Chapter 15

The Soviet Living Standard: Postwar and Later

This is, indeed, a sad chapter for me to write. Anyone who spends more than a day in the Soviet Union, not merely as a respected guest under special care, but undertaking independent activities, knows of the frightfully low living standards that exist there. It is easy to understand why the living conditions should have been very difficult after such a horrible war. Indeed, it is more than understandable, and we Germans had the least right to look down at the situation. Unfortunately, the living standard in the country is at the same pitiful level years later. This unhappy state is no longer the result of a war. Instead, it is a consequence of the economic system. The countries that have adopted the private enterprise system have long since achieved a living standard far higher than the shamefully low one in the Soviet Union.

I have already made it clear in the previous chapters that those of us working on the atomic energy project were privileged and suffered no deprivations. This chapter will, however, be devoted to the condition of the Soviet people. No one was starving when we left in 1955; however, judged by Western standards, the supply of food and other consumer goods was very low. One example is adequate. Bread was plentiful, but flour could be obtained only twice a year on the first of May and on the day of the October Revolution. Each person could then obtain one or two kilos—and this was taking place in a country that once counted grain as one of its major exports and that was famous for its Russian pies made of white flour (*pirogi*), which were found not only in the homes of the rich but were readily available to the small farmer or worker. The Soviet regime itself is responsible for the low availability of these elementary requirements at present. The Soviets cannot claim that the situation is a result of the enormous upgrading of their armaments. The United States is devoting no less attention to the development of arms and has no problem with grain. The Federal Republic of Germany had to rebuild its cities and a large part of its industry anew and nevertheless lives in plenty.

Regrettably, the situation regarding ordinary needs remains even today (1988) beyond contempt. Facilities for tourism in Moscow and Leningrad are so shabby that one would hardly be able to find counterparts in the poorest districts in the West. (One cheerful exception is the Russian ships that make cruises under the direction of Western tourist agents and can be highly recommended.) During the years that we were in the Soviet Union (1945–1955), one could not obtain a good meal in a restaurant in Moscow. Since then, however, it has become worse rather than better, which I ascribe to the inability to handle the present flood of tourists as a result of the failure of the Soviet system to grow in such a way as to accommodate them. The wares in even the best department stores are now, as previously, pitiful. I can make this statement with authority based on a recent visit to Leningrad. Reflections on these sad matters come over me again and again when I witness, in any small German town, the surplus of wares of all kinds. Poor Soviet housewife, I reach out in consolation with the hope that you do not witness this affluence and have to experience the plight of comparison.

With all of these sad declarations, I must say one word of compensation to the culinary glory of the Russian people. Within the framework of a culture of a people, one must also include along with the language, the songs, the dress, and the dances, the cultivation of the gourmet arts. Before I lose myself in enthusiasm for Russian cooking, let me ask forbearance of the reader who has only known the Russian gourmet culture from the viewpoint of a tourist. It is difficult for tourists who visit the Soviet Union to appreciate how outstanding, varied, and tasty the earlier Russian and Ukrainian cooking was. Unfortunately, only small segments of these delicious dishes are familiar in the West. In addition to borscht soup, which is well-known, there are many other very savory soups. There are many varieties of cabbage soup, vegetable soup, marvelous fish soups, special dishes of fresh water fish, the famous pies made of various doughs and various fillings (cabbage, meat, fish, rice, mushrooms)—and much more. Well-noted were the fine meals of the ordinary people that one found in the most modest homes. In the better situated circles, there were also to be found occasional Polish, Baltic, German, Jewish, Finnish, and French cuisines. Most such recipes have been almost forgotten in Soviet times. There are still, however, some of the grandmothers (the *babushkas*) who still retain the secrets of Russian cooking.

I must say something about the role of the *babushkas* and their counterpart, the *nyanyas*, in Soviet life. Because all mothers work there now, the *babushkas* have the responsibility for the entire household and the children. They are the principal support of Soviet family life. More than that, they are the preserving element of the people. If, however, one reads old Russian novels, one finds a description of another type of Russian woman who worked in a hidden manner and played an important role in the development of Russian children, particularly the children belonging to the intelligentsia. She is the Russian "nyanya," the children's caretaker who came from simple people and, while counted among the service personnel in the household, actually was considered more a member of the family, playing the role of a second mother. It was also true that, in providing personal care to her pupils, she provided a link to the life and nature of the ordinary folk. My *nyanya* in St. Petersburg, who was literate, taught me to read and write in Russian and, indeed, in a very short time without any pedagogical–psychological dislocation. Whoever had a *nyanya* thinks of her always with emotion and gratitude. Pushkin sang praises of his *nyanya* in a noble poem. He who does not appreciate *nyanyas* and *babushkas* does not know Russia.

The *nyanyas* are gone and the *babushkas* die out. I do not know whether a substitute for them will be found; however, they represented in Russian history the essential female matriarchal element, and in their passive power of resistance and patience represented an almost immovable source of strength. If Napoleon and Hitler had appreciated this strength, they would hardly have dared to cross the Russian border.[1]

Let us return to the gray reality of Soviet life. When we first entered Elektrostal at nightfall, we were surprised to see that every window in the homes was lit up. We no longer wondered about this after being there a few weeks. The explanation lay in the fact that an entire family lived in each room, and in some cases either two families or several single persons. The dearth of housing was always great in the Soviet Union and particularly so in the postwar period. During that time, there arose a story of the deserving activist who was being honored in his factory and received a picture from Stalin as a reward. However, he

---

[1]History repeats itself. Earlier, Charles XII, the eccentric warrior–king of Sweden, experienced a disaster similar to that of Napoleon and Hitler when he attempted to take Moscow in an invasion of Russia with a highly experienced army during the reign of Peter the Great (1672–1725). The climactic battle of Charles's two-year campaign took place at Poltava, southeast of Kiev, in 1709. Accidents, weather, scorched earth, bad judgment, as well as Russian valor, contributed to the defeat. The Swedish army was all but destroyed. Charles escaped capture by fleeing south to Turkey. He eventually returned to Sweden, but Russia became the dominant power in the eastern area of the Baltic Sea.

showed no great joy in accepting the present. "Aren't you pleased with the gift?" asked the director of the factory. "Where shall I hang the picture?" responded the activist. "On the wall, of course," said the director. The activist lamented, "But I live in the middle of the room." We can hope that in the intervening period, everyone has won a place on the wall to hang a picture.

The terrible shortage of housing has produced in its course an effect on the use of the Russian language. The Russian word "*kvartira*" once had the same meaning as "home." It designated a closed off living unit consisting of several rooms, including a kitchen and possibly a private toilet. In Elektrostal they used the term *kvartira* when they had a room they did not need to share with another family. One day my chauffeur said beamingly that he had finally obtained a real *kvartira*. I joined him in sharing his pleasure and asked him for details. He replied that the *kvartira* consisted of a large room in which he and his wife and two children lived in one corner, whereas his parents slept in the other.

On this occasion, I cannot deny myself the privilege of returning for the third but last time to the matter of insects. This time I will not focus on fleas, but on bugs. On one occasion I happened to ask the chauffeur mentioned above whether he had bugs in his *kvartira*. "No!" he said, "We do not. Only my parents do." Because the parents lived in the same room, I find the devoted attention of the insects to the older members of the family not only touching but also surprising. I am glad to refer this matter to the attention of the behavioral scientists and the entomologists.

The plague of insects was always a problem in Russia but was more tiresome immediately after the war. Fortunately, the situation improved a great deal after the introduction of DDT. Whereas one could regard the living bugs as the main object of hunting expeditions before DDT, present day bug hunting is focused almost entirely on bugs of the electronic variety.

Regarding the supply problem of the Soviet people, let me end by noting how interminably long a period it has been in which everything in the country has functioned badly and that only our private enterprise system, built upon motives of self-interest, produces good results. Regardless of ideals, I would rather buy items in a good store run by a self-interested individual than in a bad one run by an altruist.

Let me return to a happier theme. The Moscow subway system (Metro) is well-known because its magnificent, palatial stations. Once when the NKVD colonel, mentioned in Chapter 2, who had been a metallurgy professor, pointed out the Metro to me, I asked him whether it made sense to spend so much on a common carrier. "Please understand, the matter has educational grounds. Can you believe that anyone would spit in such a beautifully ornamented station?" I must confess that this argument was convincing. Here the old revolutionary slogan, "Palaces to the People," is truly realized, and why not? To what goal does one direct socialism in the Soviet Union, profitable or not? Were the palaces of the previous nobility profitable? We should not give up these palaces. Do not organizations such as banks and insurance companies build unprofitable business quarters? Who would say that religious bodies do not understand why they ornament their cathedrals so splendidly? The concept of profitability is not easy to come to grips with in practice, particularly in fiscal terms, because emotional factors such as prestige and the sense of beauty residing in the "customers" must be taken into account in the process. Would not our Western railroad trains be better advised to take this into account in struggling with the competition represented by individual private means of transportation, instead of allowing public transportation to become degraded?

Although the Moscow Metro must be regarded as a wonder, the lack of imagination in other fields of transportation is quite as remarkable. In contrast to the Western governments, including those that are social–democratic, the Soviets could risk more experimentation with public road transportation, particularly when designing automobiles. Nothing of the kind occurs. In all technical things they adopt imitation or, in the best cases such as rocket development, further development of Western inventions. The consequences are sometimes comical. If a Western automobile company introduces some useless gimmick, the Soviets will copy it, regarding it as "cultured." One might expect more in the way of truly new technical concepts and risks from the "First Socialist State on Earth." One cannot forget that the Soviet Union is already 70 years old. The industrial system has been forced indeed, but when will a living standard comparable to that in the Western world emerge? Where is the impulse to emulate in that field? Nothing has developed in the Soviet Union that we can really envy. After 70 years, one can rightfully raise the question: Is that all?

The reader has probably noticed that my account has the aspect of an alternating hot and cold bath: Ugly matters alternate with the beautiful. Following the hymn of infatuation with the Moscow Metro to which I added some critical comments, I turn to something ugly. I am thinking of the public toilets in the factories, bureaus, and other institutions. In many such places in the Soviet Union the human sensory organs are overwhelmed. The construction and arrangement for use of the facilities is usually different from ours. One does not sit but squat. Moreover, the units are not separated by partitions. Here is, indeed, the collective mentality, the concept of teamwork, realized in a way all its own. It is understandable that these establishments are also used as places of social and human contact. There, one can talk to colleagues without being disturbed by the boss. I am not beyond thinking that the national security service will eventually introduce listening devices. By filtering out background noises, any rashly spoken word could be captured.

I would like to mention a product having a mind of its own that emerged out of Soviet economic life—namely the Black Market, with its link to the underworld. We came into contact with this world only once. My wife and I wanted to acquire a better radio. A young German, who had previously been a war prisoner, offered to procure one. He had made contact with the underworld in Moscow while a prisoner. This happened as a consequence of the fact that the prisoners had a fine, flourishing workshop where they could repair radios and other devices. Procurement of units for repair and the subsequent distribution of their product was in the hands of the underworld. The young German went to Moscow in my automobile and looked for an acquaintance from the underworld whom he had known while in prison. The place was in a cellar, somewhere in a backyard, and was full of equipment, mainly from Germany. One could obtain almost anything through the Black Market. The currency reform that took place in Russia in the 1940s dealt the Black Marketeers a severe blow. With it, all bank notes, but not bank deposits, were devalued 90%. The Black Marketeers could not deposit their illegal money in a bank. They bought all kinds of useless things in the days before the devaluation in order to have something of possible value to show for their money. The penalty for such operations was greatly heightened in the 1960s. Nevertheless, the Black Market continues up to the present, as every tourist can see. The tourists should, however, be warned against dealing with it because they can be trapped and arrested, and receive serious penalties.

Whenever one thinks of Russia, one thinks of vodka. As a result I will close this chapter with a discussion on the consumption of vodka. Alcoholism was always widespread in Russia. Even today the problem is troublesome to deal with and energetic measures are taken against it. I did not drink much vodka, using it only for purposes of demonstrating solidarity with others or to demonstrate my ability to hold liquor. (Demonstrations of solidarity were not uncommon; demonstrations of my ability to consume were rare.) Once when I was swaying, slightly drunk, through the dreary main street of Elektrostal, it suddenly became clear that it was much easier to bear Russia if one was slightly befogged. It is always the attempt to escape from reality, whether in Russia or elsewhere, that leads to alcoholism. The monotonous landscape and the long winters in Russia contribute to this.

Immediately after the war, drunkenness understandably became fairly common. In fact, the first time I entered the management building in Elektrostal, I had to step over a drunk. He lay in the entrance way, and no one took the trouble to remove him. I recall a major in full uniform who fell into a deep muddy ditch. He was hauled out, thrown into a delivery truck, and taken somewhere to sober up. His face and his uniform, including all the decorations and indications of rank were covered uniformly with a layer of mud. As result, we referred to him later as the "Gray Major."

The drunkenness had very bad consequences in the immediate postwar period. Once I counted nine automobile wrecks in the 60-kilometer stretch between Moscow and Elektrostal. The drivers were, for the most part, former army drivers who were not familiar with the strict discipline needed on normal roads. The regime finally brought the frenzy to an end in one day by using Draconian measures, namely lifting driver's licenses. This punishment was much more effective than our useless warning signs. Moreover, this practice of the Soviet government could be recommended warmly for imitation by other governments.

It would be unfair to relate the vice associated with vodka without mentioning my own derailment in this area. I will select an example that, insofar as it had meaning, disturbed my relationship with an entire Republic of the Soviet Union. I was on a business trip with Wirths in a region of the foothills of the Urals in order to visit a factory. After taking care of the business, the head of the factory invited us to go hunting. The hunt was unsuccessful, but the process of draining a bottle of vodka was not. We were told that we were in the territory of the "Autonomous Udmurtic Republic." I was familiar with many of the autonomous republics of the Soviet Union but not with the Udmurtic. (In the meantime, I have determined that we were dealing with an ethnic group that was designated the Votyaks in old Russia.) I desired to see a typical Udmurtan. In passing through villages, several individuals who had typical Udmurtan features were pointed out to me. My vision, however, was so distorted by alcohol that instead of seeing a face, I saw no more than pink blobs. I regret very much that I cannot prove to myself that the Udmurtans have real faces instead of pink blobs. As a result, I retain to this day a guilty feeling about them (and their women whose beauty is not questioned). If fate determines that I visit the Autonomous Udmurtic Republic again, I will remain stone sober in order to retrieve what I have missed.

Chapter 16

Some Impressions of Political Life

Above all, political relationships in Soviet life show an astonishing similarity with those in Hitler's Germany. The interrelations with the private economy were different in the two countries. However, the way in which dictatorial power was used both in method and structure was similar in detail: There was the same overreaching influence of the party in relation to management and family. There was strict political registration of the youth (Hitler Youth and Komsomol). In both,the influence of the church was suppressed. There was the same influence of the party in competition for jobs and the search for work and in rewards for women who had many children ("Mother's Crosses" and "Mother Nobility"). The list could be extended indefinitely.

It is worth mentioning an incident that demonstrated to me the great attempts of the party to replace the old influence within the family by some form of collectivizing action. The goal was not to break up the family as a structural unit but to retain it as such with appropriate attention to ideology. The same had been true in Germany. One day the wife of one of our Russian engineers came to me and said that her husband had taken up with one of my women laboratory workers. She wished me to transfer the latter to some other location. I replied that this would be very difficult for me to do because I was a foreigner. I could, however, speak to the director of the laboratory. "For heavens sake, do not do that!", she responded. "The party would look into the behavior of my husband and we would be cast out of it for breaking up the family." In ways that were never made known to me, the woman in the laboratory decided on her own to switch jobs and left.

I was quite surprised when I also encountered the "Leader Concept" (*Führer Princip*) in the Soviet Union. The director of the factory once asked me on one occasion to give orders to the German workers on some matter that did not bear on our professional activities. I said that I would discuss the matter with them. "For what purpose?" he replied in astonishment. "Haven't you heard of the 'Leader Concept.'" This was not intended to be a mere play on words used by the Nazis but instead a seriously intended reference to a truly Soviet concept. For readers who know Russian, the Russian expression is *"Princip Yedi-nonatshaliya."*

At one time I had an interest in knowing whether the similarity with Hitler's regime extended to the very highest levels. On the outside, the opposite was declared very loudly, and the people had to believe it. My curiosity was satisfied at the end of 1945 on a special occasion. Some new restriction was being introduced. I went to Moscow to lodge a protest, knowing that nothing would come of it. I arrived at the office of P. J. Antropov (not to be confused with Andropov) who had been the deputy to the minister of munitions during the war; that is, to Vannikov. Later, he served as deputy and then became minister of mineral resources.

From the very start, I had good relationships with this sympathetic individual who was somewhat older than I. I believe it was mutual appreciation of old Russia that led to the formation of this special bond. I brought my protest to him. He discussed some unrelated matters in a friendly way. Then he got around to the issue at hand, during which I said, "I would be interested in knowing whether it was appreciated within your circle of management that your regime is almost identical to what ours was between 1933 and 1945." Antropov hesitated a bit and then said, "Yes, that is known." I would not have raised such a heretical question if another person had been present. One thought: Outside on the street one still sees signs bearing the slogan, "Victory of Righteousness over Fascist Germany."

The similarity between the Soviet and Hitler regimes is not unknown even in a few of the lower circles. In attempting to explain to me the nature of the NKVD, the commander of our factory in Elektrostal once said to me, "It is the same as your Gestapo."

It is astonishing that the recent dictators, who can be regarded as arising from a pendulum-like reaction to the 100-year epoch of liberalism, are almost identical in such different countries as Italy, Germany, Russia, and Japan, and that they employ such similar forms of power structure to achieve control.[1] It is almost as if they had entered into mutual agreement. This similarity was noted in folk humor. In Hitler's time the story goes that Mussolini, Hitler, and Goebbels had opened a photography shop that was very successful as a result of their cooperative effort: Mussolini developed, Hitler copied prints, and Goebbels enlarged.

The German and Russian people have much to regret because they selected Hitler and Stalin as dictators. Consider how enviously the French cherish the memory of Napoleon, or should one recall Rathenau's declaration[2] that each people experiences what it merits?

With these digressions, let me get back to the Soviet Union and to a development there that will deserve considerable watching. I refer here to the growing awakening of Russian national feeling, indeed, the tendency toward outright nationalism. This tendency was initially encouraged by Stalin during the war in order to strengthen the resistance of the Russian people. Old Russian war heroes, famous Czarist generals, and even the ancient Greek Orthodox traditions were

---

[1]Although Riehl is undoubtedly correct in emphasizing that dictatorial governments develop similar practices in exercising their powers, the basic premises upon which such governments are founded can vary significantly, giving rise to greatly different perceptions on the part of those outside of such systems. The Italian, German, and Japanese dictatorships were based on narrow concepts related to "blood and soil," whereas the Soviet government was founded at the start with the promise of freeing "enchained workers of the world" and creating a productive fraternity of them, just as Napoleon's dictatorship ultimately rested on the promise of "Liberty, Equality, Fraternity for All." All too common human greed and the desire to extend personal power gained the upper hand and controlled the course of evolution toward common patterns in such cases. I am grateful to Abraham Zaleznik for reminding me of Immanuel Kant's dictum, "The highest master should be just in himself and yet a man. The task is therefore the hardest of all; indeed its complete solution is impossible, for from such crooked timber as man is made of nothing perfectly straight can be built." (*On History*, Lewis W. Beck, Robert E. Anchor, and Emil L. Fackenheim, trans.; Bobbs-Merrill: Indianapolis, IN, 1963, pp 17–18). I am grateful to Lewis Beck for providing me with information regarding this source. I have, however, altered "crooked wood" in the translation given in the book to "crooked timber." Unfortunately, the idealism expressed initially in revolutions such as the Soviet and Communist Chinese ones may attract outside admiration, as well as devoted adherents, long after the idealistic principles have been abandoned.

[2]Riehl refers here to Walther Rathenau (1867–1922), a German statesman, industrialist and philosopher who played a major role in organizing the conservation and distribution of basic materials during World War I. He served as foreign minister in the immediate postwar democratic government and was assassinated by extreme nationalists for supporting the terms of the Treaty of Versailles and seeking close relations with the Soviet Union, the other major "outcast" nation of the period.

brought forth to create a national euphoria in support of the war. Stalin himself checked, oversaw, and modified each historical book and film, both during and after the war in order to give emphasis to the heroic elements of Russian war history—for example, a book on the Russo–Japanese War and a corresponding film. I recall a naval war film relating to the time of the Russo–Japanese War in which a heroic Russian admiral played the principal role. The film was historically accurate. The well-attended admiral, a grand seigneur par excellence, acted in a distinguished manner in his enormous, luxurious cabin. He made such a striking human and military impression that one might, after seeing it, ask whether it really was necessary for the master revolutionaries to kill off the Russian upper layers that contained such splendid men. If the film had been shown in Czarist times, the director would have been ennobled and the "Left" would have accused it of being chauvinistic.

Following the victorious ending of the "Second Great War of the Fatherland," the wave of nationalism rose understandably higher. The raising of the Soviet Union to the status of a world power, the rapid growth in military strength, and the consequences of Sputnik all advanced and increased the national euphoria. I shall relate a typical example of this posture. As I went into a phonograph record shop in Moscow at the end of the 1940s, a well-dressed man about 50 years of age entered and enquired about Beethoven recordings. He regarded himself as an enthusiastic admirer of Beethoven. "Do you know," said the saleswoman, "Beethoven is not a German to me but a Russian!" I recognized in this statement tones that I recalled from old Russian times. Alongside the Russian feelings of inferiority of that time, which rested mainly on social and technical matters related to their level of civilization, there was also present a messianically colored national pride. This posture is not unknown to us Germans. "The German essence nourishes the world" (*"Am Deutschen Wesen die Welt Genesen"*). Weren't the Germans members of a "higher race"? Such nationalistic excrescences always have their roots in some kind of hidden inferiority complex. Let it be said: When healthy, justifiable national pride is transformed into a messianic form, the good Russian, or for that matter good non-Russian, individual will no longer exist and will be replaced by a highly dangerous, far-reaching reformer–disciple.

One cannot allow this Russian nationalist tendency that lies behind the facade of a self-congratulatory mood to go ignored.[3] In other respects, the Russians have much justification to possess a strong

---

[3]The politically ambitious Vladimir Zhirinovsky is attempting to exploit this current of expansionist nationalism in the Soviet Union at the present time.

sense of nationality. Their country has existed for more than 1000 years. It has, in part, a proud history and has an immense ability to assimilate others. Russian language, music, and literature alone provide justification for the Russians to be proud of their country.

The Stalin cult was blossoming richly when we first came to the Soviet Union. He was the Genial Commander, the Benevolent Father of the People, the Wise, Great Scholar. Hitler, who unpretentiously never allowed himself to be designated as anything more than "the Greatest Commander of All Time" (Gröfaz), had cause for envy. The homage paid to Stalin took grotesque forms. As a Hero of Soviet Work and bearer of the Stalin Prize, I was always invited to the great celebrations on May 1 and on the anniversary of the October Revolution, with the privilege of sitting at the president's table. At the very beginning, it was announced that a motion would be made on the part of colleagues of Stalin to send a congratulatory telegram. There was then an extended cheer. The telegram, which was yards long and had been prepared earlier, was then read and was followed by additional acclamation. Similar events took place in factories and other institutions in the country.

In the celebration of Stalin's seventieth birthday, a hurricane of applause broke out that was limited only through physical exhaustion whenever the words "Stalin" or "Josif Vissarionovich" were mentioned. When Stalin died, there was general mourning and hysterical expressions of grief at a level that can hardly be found in human history. For example, in the student dormitory where my daughter lived with eight other women students, the young women threw themselves on their beds and blubbered pitifully. Through mass psychosis, my daughter howled along. A wake was held in Agudseri on the Black Sea. The Soviet functionaries begged me to say something in the name of the German specialists. I took care of this responsibility by giving a speech that had the following tenor: "We have worked very hard together for many years, shoulder to shoulder. Through this we have come closer together. Your joy is our joy, your sorrow is our sorrow." Thus I gave a respectful presentation without compromising myself. The other speakers slimed themselves with heart-rending mournful speeches. One speaker broke down sobbing and had to leave the podium. My family and I were no longer in the Soviet Union when Khruschev denounced the great Stalin as a master criminal.

I would like to mention another peculiar issue related to public life in the Soviet Union that I never really understood. While in Berlin, shortly before being transferred to the Soviet Union, a well-intentioned young NKVD officer said to me that in the event that some failure occurred in our work I should not be shy to shoulder the blame fully and at once, and be certain that all the details of the matter were revealed. I did not grasp his advice completely, but I did make a definite note of it. I recalled it when Colonel Uralets, the fine project head in the Urals, told me in conversation what to do when one encounters blame for some reason. He related this in terms of an event in his own life. At a regional party gathering, a young woman and he were attracted to one another and a seduction followed. He then learned that she was still a minor. He did not respond with indignation as one of us would have done. Instead he gave a talk to the assembled group saying, "I seduced Comrade N. who was not of age. Comrades, seduction is a serious offense. A crime of this type is not tolerated. I know that the communist vigilance and dedication of Comrade N. are highly treasured. I regard her, in fact, as a model and can recommend her to you." He continued in this style until he came to a point where he said he had not intended to seduce an under-age individual. He closed his speech by giving additional thanks to Comrade N. The procedure was a complete success. Comrade N. was intentionally and consciously in the middle of the gathering in complete awareness of what was going on. The audience appreciated the spicy situation, and everything was placed in proper order. The colonel related this not as a single unusual exception but rather as a thoroughly normal type of event. I do not know from what psychological roots of human behavior this practice stems. Is it typically Russian? or Soviet? Does one have to regard it in the same vein as Dostoyevsky's sense of pleasure in knowledge of his guilt, or with the pleasure in the confession, or spiritual exhibitonism, or to masochism as one observes elsewhere as well. For lack of confidence, I must limit myself to giving an account of the phenomenon.[4]

Many of us specialists were involved in the functions of the Communist Party. My own experiences in this area were sporadic. Because party officials had no obvious administrative duties, their influence was indeterminate. The party worked in the background as a control and surveillance organ in which the supervisory functions were focused on the ideological, occupational, and moralistic welfare of the party members. In Elektrostal the local party officials were present at all important discussions and conferences but rarely entered into them. Any censorship they might exert was primarily taken in connection with their reports to superiors, both in accordance with instructions imparted to them from higher party regulations and from the recommendations of local party gatherings. In the

---

[4]The social practice of "confession" described by Riehl may date back to the time of Peter the Great who would forgive individuals for lapses in behavior if they made a clear and full confession—provided the incident did not involve danger to the state or violate some other high principle.

main, the role of the party resembled that of the church in the European Middle Ages. Exclusion from the party had much the same catastrophic effect on the recipient who was affected as excommunication from the church had in old times. Our factory director made frequent visits to Moscow for meetings of the Central Committee (ZK). He was a very ambitious man and clearly strove to obtain a position as a deputy minister, which required being regarded highly by the party. Directors of large and important factories often rose to posts as deputies to ministries involving professional specialist areas. As is well-known, there are many ministries for all possible branches of industry, commerce, and transportation. The "classical" ministers, such as those for interior and foreign affairs and the like, were, however, selected on the basis of political influence. Beria was the only one of this type I came to know. On one occasion, I had a brief opportunity to speak in the Kremlin to Shvernik (see photograph on page 104), the chairman of the Presidium and the highest Soviet official. He presented me with the Golden Star of the Hero of Socialist Work and the Order of Lenin. He was a friendly, cultivated man who at one time had the same position later held by Podgorny, that of a representative of the Soviet States.

When portraying the behavior of the Soviet population, a question arises concerning the extent to which religiosity currently plays a role in the Soviet Union, or could play in the future. The tourists are taken to see overflowing churches and frequently conclude that there is great interest in religion in present-day Russia. Such a conclusion is, however, not justified; most of the countless Russian churches are closed and what the tourists see is a concentration of the faithful in the few which are still available. Of greater importance is, perhaps, the fact that many young people have joined the believers, yet one cannot conclude from this that religion has had any influence on the trend of political development.

I believe, in following up on this discussion, that one should primarily attempt to keep in mind the value ideologies possess when viewed from the position of the politician—independent of their ethical value, good or bad. The primitive cult of gods, the ethical foundation of newly founded religions, the monarchal principal, nationalism, socialism—all have functioned or function as extended arms of power when viewed politically. They have all exerted or exert a state-preserving function. Or expressed primitively, it is easier to control people by the bridle when one impregnates them with a halfway believable concept than would be the case if one had to handle each one individually with a whip. The Greek-Orthodox religion played the fundamental role as a state-preserving ideology in the old days. One fought and sacrificed "for the Faith, the Czar, and the Fatherland." The faith stood in first place, even before the Czar! One fought with the "infidel" Tatars or Turks, with the Poles who had a different form of faith, but not with the Greek-Orthodox Georgians.[5]

Anyone who knew Russia could confirm that the Russians had a strong tendency, along with the ability, to adopt a concept and sacrifice for it. It was religion in the old Russia that occupied this position. In more recent times, other concepts provided the stimulus for enthusiasm. The Russian revolutionists did not carry out their assassinations from secure hiding places like present-day terrorists. They threw most of the bombs intended for sacrificial killing in the open and thereby sacrificed themselves along with their victims. In following an ideology, the Russians are more closely progammed to symbols than most other people. When in old Russia one witnessed the ardor with which many of the faithful crossed themselves, one could believe that the crossing was not merely a symbol but the actual content of their belief. This joy in following and adhering to symbols of all types is perhaps rooted deeply in the Russian character. This accounts for the pronounced love of the Russians for the ringing of bells, for the expressiveness of the Russian language, for the tendency to pronounce many words or names with a sonorous voice or with sing-song overtones, for their inclination to generate picturesque formulations, and for their variable turn of mind.

If it were only a matter of the return of Russian religiosity in the future, I would express my personal opinion as follows on the basis of the knowledge of the old and the new Russia: One should look upon the religious tendency of the Russian people as one expression of a more general tendency to turn to some form of ideology in an inspired way and to serve it. Of concern to us at present is the fact that the Soviet leadership has become restless and has determined to shift this capacity to become inspired into a new direction of its choice in which the socialist ideology would take on obvious nationalistic features. The earlier, prominently expressed Russian capacity for religious behavior has been placed in the service of another ideological goal. The new behavior is so tightly programmed in the Russians, so "cemented" through familiarity, that the acceptance of new symbols involving an abrupt change in course cannot be anticipated in the near future.

The reorientation of the Soviet people into the new "communist" ideology was not, as it happens, the product of open discussion and of

---

[5]During World War I the Turks became fearful, perhaps correctly, that the Christian Armenians would not resist the invasion of Eastern Turkey by the overtly Christian Russians. This led to the brutal slaughter of a large segment of the Armenian population by the Turks. The Armenians were later befriended by the leaders of the Soviet Union and permitted to form a separate republic.

free and convincing intellectual activity. It was the product of dictatorial pressure, the hammering in of new solutions that were drawn up by the authorities, just as was the case for us in Germany. The mass of Russian people hardly had time,[6] following the fall of the Czars, to become accustomed to a free regime and to become fully devoted in their behavior to the intermediate leadership (the Kerensky regime). The October Revolution took place just a few months after the Czarist system was disrupted and only a mere 50 years after the freeing of the serfs.

If one ignores the intellectual stratum of the society, one notes that the population experienced an essentially seamless transition from one type of authoritarian government to another. I was often astonished at the readiness with which the same population was prepared to subordinate itself. I often heard an expression that went back to Czarist times: "The authorities see more" (*natshalstvu vidneye*). In one instance, we saw an example of this behavior that could not be distinguished from that of a serf. While in the Urals my wife gave our cleaning woman, a poor simple person, a used topcoat that was in good condition. The incident took place in the kitchen of our house while I was having breakfast in the dining room. I heard the grateful stammering of the cleaning woman and then a soft cry from my wife. The cleaning woman ran into the dining room, grabbed my legs, and kissed my feet. My wife ran into the room and cried in horror, "She kissed your feet!" She was very indignant that I was little concerned and that I calmly continued eating my breakfast. Here were human beings from three different worlds, the Central European, the simple woman from one of the most backward Russian provinces, and the born Petersburger who had never managed to have his feet kissed previously, but on whom the old Russia had left such a deep mark that he thought nothing of it.

This spirit of subordination was not restricted to the tutorial regime but also appeared in the economic structure of the country. Our factory, for example, contained all forms of sub-organizations that were needed for its complete operation. It had, for example, a manufacturing staff, a food staff, schools, a culture club, living accommodations, and so forth, more or less under the directorship of the factory. As a result,

---

[6]In February of 1917, during World War I, the moderate Russian socialists overthrew the Czarist government and established what was intended to be a democratic form of socialism. One of the leaders in this activity was Aleksandr F. Kerensky, who became Prime Minister in July with the intention of continuing Russia's role in World War I. The arrival of Lenin and the more violent Communist revolution of October 1917 brought the liberal so-called Kerensky regime to an end. The Kerensky regime spared the lives of the royal family, whereas the Communists did not. Kerensky, incidentally, escaped from the Communists and eventually emigrated to the United States.

А. Ф. Керенскій

Alexander F. Kerensky, a principal leader in the attempt to establish a democratic government in Russia during the spring of 1917. He and his fellow democratic revolutionaries deposed the Czar and his family, confining them in dignity and without harm. He was supported by the Western Allies with the assurance that he would keep Russia active in World War I, which was now very unpopular in the country. As a result, the Germans transported Lenin from Switzerland, where he was in exile, to the Russian-Finnish border with the promise that he would withdraw Russia from the war if his revolution succeeded. Thus began the far-reaching Soviet communist revolution in October of 1917. (Courtesy of the Russian Pictorial Collection of the Hoover Institution Archives.)

someone working in the factory not only had to follow the directives of those who ran his or her immediate service in private matters, but also had to follow the will of the factory director. As a result, any attempt to transfer to another factory or place could involve serious difficulties. Actually, such a transfer from our special factory was most unlikely because we were involved in secret work. The overall situation had some of the aspects that were common in bygone periods when the worker could be held in a form of slavery-like bondage resembling the ties that linked the peasants to the lord of the manor, under whose influence they fell. I frequently experienced cases in which the factory director used the familiar form of speech (the equivalent in German of "*Du*" form instead of the "*Sie*" form) in criticizing an employee, while the latter stood silently in a respectful stance. (This was not the practice of the "sympathetic general" I mentioned earlier who was initially in charge of our factory and who exhibited the best human qualities when he had to complain about something to an employee.) Another example: On one occasion the football team of our factory lost a game that was played against the team of another factory. In all earnestness the director threatened to fire the members of the team from their factory jobs as retribution!

Inasmuch as I feel required to use the word "slavery" in describing the position of the Soviet worker, I would like to delve further into this theme. Even though one can assume that the position of the worker in his workplace has improved in the intervening period, the fact remains that the state, in keeping with its role as the legislative and executive authority, is also in the position of being the only employer. One may ask whether that does not conform to the basic concept of slavery. Because the state is controlled by a relatively small group of individuals who, in turn, observe the strictest ideological rules in their orientation, is it not possible to ask: Is a new period equivalent to the Middle Ages on its way? Was the liberal epoch only an episode in the history of mankind? In fact, many of the developments that we are experiencing in the West encourage this line of thought, even though they take place against different backgrounds.

We are experiencing the effect of technology overrunning its banks, which makes us slaves of our machines and computers. Are the days of freedom numbered? When will the two halves of the vise, which is being constructed by technology on the one hand and government officials on the other, snap closed? Has the liberal way of thinking, which we in Europe have experienced since the beginning of the century, initally with the permission of the tired monarchs, been merely a luxury? Those of us who are older remember those monarchs very well. They were tolerant while not realizing that they stood on feet of clay. The thought of these monarchs reminds one of many lenient older men who muse on the fact that most of the mistakes they made during their lifetimes they actually projected while young. One calls such musing the wisdom of the aged. This wisdom is somewhat good but it provides no defense against approaching death. However that may be, and as long as one is not yet blocked in, may I echo the cry: "*Vive la liberté!*"

Let me return to my theme and, indeed, to the important question of whether we are at the beginning of a period when we may expect the Soviet regime to provide more freedom. During the years I was in the Soviet Union such estimates were only potentially conceivable. We were, however, most conscious of the great desire of the people for more comfort, more variety, and more color. The following is a typical example. We Germans were allowed to receive no foreign newspapers other than those from East Germany. Their contents were not entirely useless because we had learned to read between the lines during Hitler's era. Our pleasure was greatly increased when one of our group succeeded in receiving copies of a Western magazine, "*Konstanza,*"—a woman's fashion journal. The Soviet women literally tore the magazine out of our hands. Even complete party-liners risked a stolen glance at the pretty pictures in the journal. There was an elementary longing toward such beauty and color that could not be remedied by other things.

More important, and easier to comprehend, however, were the perceptions that I was able to form later on in Germany when I met with Soviet scientists, particularly the younger ones. By the beginning of the 1970s, their behavior had changed completely. They became much more open, talked more freely, and no longer were bound to old attitudes and words. I could, without embarrassment, relate historical accounts of the type presented in this book to them without causing them to feel even a bit ill at ease. They listened with most intense interest to the history of early events in their country—events that were withheld from them at home or were deliberately presented in a highly distorted way. Moreover, they were good sports with regard to my presentation, showing no resentment because they realized that no criticism or mockery of their people was intended. Let us hope that the improvement of political developments in that great country continue—and that its neighbors also grasp the idea.

Commentary Note

One of the visitors from the Soviet Union with whom Riehl developed a warm friendship after returning to Germany is Zhores I. Alferov, currently vice president of the Russian Academy of Sciences and director of the Ioffe Physico-Technical Institute in St. Petersburg. In a letter to me dated February 14, 1995, Alferov makes the following informative comments:

It is very interesting for me that you are working on the preparation of the book which is connected with the activities of Dr. Nikolaus Riehl. First time I met Prof. Riehl in the beginning of 1966 during my first visit to Germany at his laboratory in Munich Technical University. He invited me to his house on Sunday and told me his story during 14 hours from 11 a.m. up to 1 a.m. Later I met him in Germany and USA, and had two hours telephone conversation last time in 1988 during my very short visit to Germany. He was a very interesting person and I keep in my memory practically everything what he told me.

He always refused to write the memoirs and explained this to me in the following way: "When I am telling my story, people from West and East understand me well. Had it been written, both sides would not have understood me perfectly." His short memoirs in German which he sent me I could not estimate exactly due to my bad knowledge of German, but I am afraid that he lost in the written memoirs a lot of what he delivered on oral presentation. In this type of conversation he was quite an artist. In spite of big difference in age and may be in some approaches to different life and political problems we became friends and every possibility of communication with him I appreciated very much. He was born in S. Petersburg and shortly after the revolution graduated from Peterschule at the Nevsky prospect. He was speaking excellent Russian with an old Petersburg's pronunciation. Now just a few people in our country are still alive who were working together with him in 1945–1955. It would be very interesting for me to be familiar with your book.

Chapter 17

The Lysenko Affair

This is a gloomy chapter in the history of the relationship between the Soviet government and science. At the end of the 1940s a campaign was kindled against the work of Mendel, Weisman, and Morgan in the newly developing field of genetics. The principal actor functioning in the campaign was the agronomist, T. D. Lysenko.[1] It is difficult to determine in detail just which individuals and motives were behind the creation of this campaign. The motives were primarily political in nature and the individual primarily responsible must have been Stalin himself. Without his permission, in the last analysis, no official position

---

[1]The sordid history of Lysenkoism is presented in much detail in Zhores A. Medvedev's *The Rise and Fall of T. D. Lysenko* (Columbia University Press: New York, 1969), David Joravsky's *The Lysenko Affair* (Harvard University Press: Cambridge, MA, 1970), and Mark Popovsky's *The Vavilov Affair* (Archon Books, The Shoe String Press: Hamden, CT, 1984). Popovsky spent a critical part of his career in the Soviet Union investigating the life of Nikolai Vavilov as well as the circumstances under which he died in prison early in 1943. He concluded that the high authorities such as Beria who might have been willing to release him as a useful prisoner involved in agricultural research and development were prevented from doing so at the insistence of Lysenko, undoubtedly backed by Stalin. As a result, he was allowed to starve to death over an extended period of perhaps a year and a half. Popovsky, who had previously been a highly regarded Soviet writer, was exiled to Western Europe for his research in behalf of N. Vavilov, as was Medvedev for similar reasons. I had the privilege of serving for several years on an advisory committee with the American geneticist Herman J. Muller, who had spent several years with Vavilov in the 1930s. Muller indicated clearly that the period was one of the high points in his own career.

Incidentally, Popovsky describes the decay and turmoil that developed in St. Petersburg in 1917–1918—the disintegration that made it desirable for the Riehl family to leave.

could be taken with respect to any issue regarding a branch of science or art. Each new novel, each new film was presented to him for judgment and either accepted or rejected. Often a new version was prepared with his permission if the original did not suit his viewpoint. This type of exertion of influence was never kept silent but was often made public in order to set future authors on the "proper path." There rarely

Trofim D. Lysenko, an energetic plant breeder who came to the attention of Nikolai Vavilov in the 1930s and was given praise and recognition for his industry. Vavilov probably hoped that Lysenko could be guided into pathways of good science. Although this special support may have helped Lysenko gain national recognition and power, which he grossly misused with the help first of Stalin and later of Khrushchev, it is Zhores Medvedev's view that Lysenko's peasant origins were his greatest asset as far as the Soviet leaders were concerned, once he became noted. Stalin used wartime conditions as an opportunity to liquidate many potentially dissident intellectuals and individuals of upper-class descent. (Courtesy of Zhores Medvedev.)

was any involvement on the part of Stalin or the party when scientific issues were under discussion. When it did occur, however, he was unquestionably involved, for was he not only a great statesman and field commander but also a great scholar.

It appears that the power and the enduring adulation induce a feeling of infallibility in dictators and that they make decisions as a matter of faith on topics about which they have no understanding. It is even more unfortunate when they really think they understood something about a subject. However, it is not only dictators that can do harm by interfering in this way. It is well-known in industry, for example, that it is usually better when the executive has a good knowledge of people and a healthy sense of humanity, as well as other qualities of leadership, and does not intrude as a scientist or engineer when special problems are to be solved. In connection with this, the distinguished German orchestra conductor, Furtwaengler, made a cogent comment. Someone asked him during the Hitler era how it happened that he could pursue his art without any disruptive interference from Hitler. He answered, "It lies in the fact that Hitler did not play any instrument. But think of how difficult it would have been for me if he had played the mouth harmonica!"

The offical basis for Lysenko's campaign lay in the claim that the form of genetics being attacked was an idealistic–metaphysical form of bourgeois science that had no relation to agricultural practice and that it would hinder the development of Soviet agriculture. I learned also of another reason from announcements made that the Mendelian form of genetics provided a basis for racism in the sense that it negated the possiblity of breeding based on acquired characteristics in the Lamarkian sense. The very productive Russian agrobiologist, Y. V. Michurin, who died in 1935, was declared after his death to be the father figure of the Lysenko form of heredity. Russian biology and breeding practices, in accordance with Lysenko's concepts, were exalted and praised. A nationalistic tendency was unmistakably contained in this endeavor, as was true in many other declarations of Soviet propaganda. The Lysenko campaign had a completely tragic effect on Russian biologists and geneticists because Russia had produced a large spectrum of very distinguished investigators who had made great contributions to the advancement of genetics following the concepts of Mendel, Weisman, and Morgan. Among them were N. K. Koltsov, N. I. Vavilov , N. P. Dubinin, N. V. Timofeyev-Ressovsky. N. I. Vavilov, who died in banishment in 1942, was a brother of the noted physicist S. I. Vavilov, mentioned in Chapter 8, who served as president of the Soviet Academy of Sciences in the 1940s (see photograph on page 130). N. I. Vavilov was "rehabilitated" in 1960 with the downfall of Lysenkoism. During the course of the repression many geneticists were compelled to give self-accusatory speeches condemning their views, recognizing their guilt, and vowing improvement.

Although I am not a biologist, the Lysenko affair did have its effect on me, although only obliquely. Near the end of the 1940s I had written a book on the radiation transfer of quantum energy (migration of energy) in both inanimate and biological materials. The biological portion of the book was taken directly from the work of Timofeyev-Ressovsky, K. G. Zimmer, and M. Delbrück on radiation-induced mutations. I wrote the book in German initially but then translated it into Russian. Zimmer also translated it into English. A distinguished state-owned Soviet press decided to publish the book. The publisher provided me with the services of a young Soviet biophysicist to help prepare the manuscript for publication. Acting under the recommendation of others, the young man inserted the name of N. K. Koltsov as "genius loci." I accepted this gladly because I knew the work of Koltsov, who could be considered as one of the founders of modern molecular biology.

The book went to press, was bound, and was made known to all book distributors. When I made a visit to the press, one of the staff showed me the finished author's exemplars of the book. Had I a briefcase with me, I would have taken the copies along. However, without the briefcase, I would not have been able to take the books out of the publishing house without a special pass. The pass would have had to have been signed by the editor-in-chief, but he was away at the moment. I said that I would pick them up the next time I came to Moscow, because I made frequent visits to Moscow from Elektrostal. I returned to the publisher a week later with a briefcase. On entering the building, I was requested to see the editor-in-chief. Because the first clouds of Lysenkoism were gathering in the genetic heavens in 1948, I had bad forebodings. The editor-in-chief informed me that the entire biological side of the book would have to be rewritten. What was particularly bad was the fact that I had cited Timofeyev-Ressovsky in favorable terms and praised Koltsov. Several months earlier I had been able to glorify Koltsov, but now the same individual was a bone of contention. I do not know whether to laugh or cry over such typical Soviet changes in direction. With disarmed, trusting, open eyes and inner convictions, the Soviet people, on a sign from above, are required to believe today the opposite of what they believed yesterday. (I observed such behavior frequently, for example, at the time of the liquidation of Beria.) The decades-long period of education leading to the loss of character, decline in sincerity, and crawling subservience toward the authorities, has had its full effect. Have generations of freedom-loving Russians fought and suffered for this?

And so it went with three books that were written in three languages with much industry. They remained held in limbo. None of the

Nikolai I. Vavilov, the brilliant geneticist who was persecuted by Stalin and Lysenko in spite of his outstanding contributions to agricultural science. One of his many enlightening insights into the field of plant science was to propose that the most likely geographical site of origin of a widely domesticated species is the place where it has a number of closely related wild species. He died of starvation in prison in 1943, undoubtedly with Stalin's knowledge. N. I. Vavilov was a man of enormous integrity and courage, willing to sacrifice his life in defense of good science if need be. His courage and moral strength were on a par with those of Andrei Sakharov. This photograph was probably taken about 1930. The tragic end of his career is enhanced to a substantial degree by the fact that at the height of his justly deserved fame Vavilov, with all good will, gave enthusiastic public support to the early energetic work of Lysenko on plant breeding. (Courtesy of Zhores Medvedev.)

authors were able to retain a copy as a memoir. Later I learned that some Soviet scientists had obtained a number of copies of my book by some secret means. I found it cited later in a Russian publication. The fact that a book that has never appeared is cited has its comical aspect. A Soviet scientist whom I met later at a conference outside the Soviet Union expressed much praise for the book.

It did not bother me a great deal that the book was not published. I found a few years later that some of the biological and biochemical phenomena that I had tried to explain in the book could be explained better on other grounds. Thus, my one-time love for biology can be regarded as an "unlucky love." It was not, however, entirely unlucky because the work encouraged me to become more deeply involved in the subject and undertake further investigations that eventually led to more interesting results. For example, they led to the study of electro-conductivity of protons in ice and in many important biological materials (such as protonic semiconductors).

More grave were the consequences experienced by my colleagues N. V. Timofeyev-Ressovsky and K. G. Zimmer as a result of the Lysenko affair. At the end of the war they had written *On the Staircase Principle in Biology*, which contained a comprehensive summary of their work and that of others on radiation-induced gene mutations and related areas. The book was published in Leipzig in 1947. In 1948 it became a victim of the Lysenko campaign and was put on the forbidden list. The East Germans promptly followed the example of Big Brother. I sense that my colleague, Zimmer, who was in the Soviet Union at that time, was exceedingly downcast because the book represented the product of many years of his life's work.

Because our institute in Sungul was deeply involved in biology, the influence of the Lysenko affair was felt keenly. In fact, we were no longer allowed to carry out genetic research. Moreover, the spirit of Lysenko escalated and had its effect in other areas of science. The party extended the underlying constraint and placed other areas of science under its purview, including the foundations of chemistry and physics, which it felt needed revision. That was carried out with the ever-growing Soviet tendency to emphasize nationalism. Every questionable Soviet "discovery" was seized upon and exaggerated. Many of these discoveries bordered on the charlatan. All new theories that came from the West were denigrated for their idealistic–capitalistic content and evaluated accordingly.

For reasons that were not at all well understood by me, the Soviet ideology adopted a particularly negative view of the resonance theory of chemical binding. Perhaps it was the lack of their ability to express the underlying concepts in a concrete manner that raised their wrath. In any case, a thorough nationwide campaign against the use of the theory was instituted. This extended to the Sungul Institute where a Soviet chemist was compelled to give a negative critical lecture on the resonance theory, even though the work of the institute had nothing to do with it. The party influence was called forth in earnest whenever a theory contained philosophical elements or when the party philosophers believed it contained philosophical elements, as best they could ascertain. The well-known theoretical physicist Blokhinsev had once written a book on quantum mechanics. The somewhat philosophical introduction to the book displeased the party, and poor Blokhinsev had to rewrite the introduction seven times.

In Chapter 16 I pointed out many similarities between the Hitler and Stalin regimes. I can add another here. The Nazi party also sought to get involved with science and to elevate the concept of "German physics" in opposition to "Jewish physics." It resounds to the honor of German physicists that "German physics" could not create any damage and finally vanished on its own as an absurdity.[2]

---

[2]Two scientists who supported Hitler's government very strongly and who hoped to promote so-called "German science," in contrast to "Jewish science," were Philipp E. A. von Lenard and Johannes Stark. They were very critical of Heisenberg and denounced him publicly for accepting the theory of relativity. Both carried the prestige of Nobel Prizes for earlier distinguished work. Perhaps their nearest analog in the Soviet Union would be Lysenko, although the latter actually had more political power because he was immediately backed by Stalin. The great majority of established, tenured, academic scientists in Germany turned inward toward their professions in order to avoid political involvement.

Chapter 18
Conclusion[1]

When I look back on what I have written, I find in it much humor that is based on episodic, anecdotal, and political matters. I believe that this is necessary for the representation of reality. The episodes one experiences personally are closer to reality than any generalizations, and one can hardly imagine the political humor encountered while living under a dictatorship. The humor that emerges is somewhat like the ray of sunshine that may occur in the cell of a prisoner. I may ask myself, however, whether it is also possible to add some earnest professional generalizations to this. In fact, the reader may well ask whether I have something to say regarding the future course of events in the Soviet Union as a result of my intimate experience with Russian life, both early on and in the Soviet period. The single truly honest response from me is, "I do not know." It is difficult enough to comprehend the past. How much more difficult to extrapolate to the future! Even the most experienced politicians and economists have turned out to be false prophets. One can hardly place more credence upon their prognoses than one can upon the conclusions fortune tellers derive from examining tea leaves. Even the present-day futurologists who deal mainly with statistical–quantitative methods find it difficult to look into the future because of the intrusion of very important emotional factors that they cannot include in their calculations. The practice of futurology is, perhaps, an occupation that will always have a bad conscience for this reason. Their activity resembles that of a merchant who is trying to price a ware that may possibly be available in the future.

On the other hand, I would not like the reader to believe that I have not given thought to the future of the Soviet Union. The prognosis is not

---

[1]Although in one sense this chapter (which was obviously written before 1990) is somewhat out of date, it must be admitted that the future of Russia is still very uncertain. The transition from a rigid form of communism to a free enterprise republic is not a straightforward one, as Riehl correctly surmised.

difficult if one contemplates the prospect of another world war. For, in this case, we can repeat the comment of Albert Einstein regarding the weapons that might be used in the next world war. He said, "I do not know what weapons will be used in World War III, but I do know which will be used in World War IV, namely the spear and club." It is, however, worthwhile to offer some thoughts if world peace is to be maintained. Of particular importance is the question of whether or not a fundamental change in the economic system in the Soviet Union is still possible. I would respond to this with a plain "no." At most it is possible that there will be some loosening that permits a limited amount of private enterprise of the type that has been seen in Hungary and Yugoslavia.[2]

In other respects, the government-centered capitalistic structure is so cemented and the associated methods are so well established, that a full transformation is neither technically nor psychologically possible. Preservation of the present system might seem to imply that the living standards of the Soviet people can hardly rise. This, however, is not necessarily so and could be regarded at most only a result of wishful thinking. Where is it written that the living standard there cannot rise in due course of time? No one is starving there, and the Soviet citizen is capable of surviving for a very long time under conditions that do not vary. Moreover, nonprivate enterprises are capable of functioning. Even in old Germany, when we still had a faultlessly functioning railroad system, it was said that there were three enterprises not in private hands that functioned well: the railroads, the Prussian army, and the Catholic Church.

Regarding the broader question as to whether I believe that a real liberalization of the Soviet system is possible, I would like to express less skepticism. At the end of an earlier chapter I mentioned several incidents that indicated some generally happy changes in this direction. Also, it is significant that many of the dissidents are allowed to leave and that an individual such as Sakharov is permitted to play a significant political role. Thirty years ago this was inconceivable. If the liberalization continues, one can express the earnest hope that in a few decades a degree of freedom comparable to that one experienced in the days of the last Czars will be attained. Anyone who knew those times can affirm that it would represent a true step forward.

---

[2] It is widely known that once the bonds on a population are loosened to a degree, strong internal pressure develops to achieve even greater freedom. This is the well-known phenomenon of "rising expectations." Apparently, both Riehl and Gorbachev thought that either governmentally instituted restraining forces or common judgment on the part of the people would serve as a continuing restraint. Mainland China faces this problem at the present time.

Enough of futurology, criticism, and of rationalistic considerations. I am at the end of my story. A quiet sense of sadness comes over me, as if I had relived a part of my life for the second time. Once again my vision wanders over the wide Russian land and to the people there. In their behalf I would like to express my feeling and close with the words, "May you be spared further sorrow, dear Mother Russia!"

Bibliography

Barwich, H.; Barwich, E. *Das Rote Atom*. Scherz Verlag: Munich, Germany, 1967; Fischer Bücherei: Stuttgart, Germany, 1970.

Golowin, J. N. *J. W. Kurtschatow*. Leipzig: Urania Verlag: Leipzig, Germany, 1976 (in German, translated from the Russian).

Irving, D. *The Virus House*. W. Kimber: London, 1967.

Kurowski, F. *Allierte Jagd auf deutsche Wissenschaftler*, Verlag Kristall bei Langen Müller: Munich, Germany, 1982.

Medvedev, Z. *The Nuclear Disaster in the Urals*. Norton: New York, 1979.

Index

211

Frederick Seitz